THE ECONOMICS
OF TOURISM

M. Thea Sinclair and Mike Stabler

London and New York

First published 1997
by Routledge
11 New Fetter Lane, London EC4P 4EE

Simultaneously published in the USA and Canada
by Routledge
29 West 35th Street, New York, NY 10001

© 1997 M. Thea Sinclair and Mike Stabler

Typeset in Garamond by
J&L Composition Ltd, Filey, North Yorkshire
Printed in Great Britain by
Creative Print and Design (Wales), Ebbw Vale

British Library Cataloguing in Publication Data
A catalogue record for this book is available from the British
Library

Library of Congress Cataloguing in Publication Data
A catalogue record for this title has been requested

ISBN 0–415–08523–3 (hbk)
ISBN 0–415–17076–1 (pbk)

ROUTLEDGE ADVANCES IN TOURISM
Series editor: Brian Goodall

THE ECONOMICS OF TOURISM

Tourism is one of the world's most important activities, involving millions of people, vast sums of money and generating employment in developing and industrialized countries. Yet many aspects of tourism have been ignored.

This book makes a key contribution from an economic standpoint to the understanding of tourism. Examining such issues as demand for tourism, how tourism firms operate in national and global contexts and the effects of tourism on destination areas, the authors explain how economic concepts and techniques can be applied to the subject.

Particular attention is paid to the importance of market failure with reference to identifying the environmental implications of tourism and the means of pursuing the sustainability of both tourism and the resources on which it depends.

The Economics of Tourism presents new insights into the intricacies of tourism demand, firms and markets, their global interrelations and the fundamental contribution of the environment to tourism activities, to offer an accessible, interdisciplinary analysis of the interwoven fields of tourism and economics.

M. Thea Sinclair is Senior Lecturer in Economics and Director of the Tourism Research Centre at the University of Kent.

Mike Stabler is Visiting Fellow in the Centre for Spatial and Real Estate Economics at the University of Reading and joint Director of the Tourism Research and Policy Unit.

Routledge Advances in Tourism Series
Edited by Brian Goodall, University of Reading

To our families

CONTENTS

FIGURES AND TABLES

FIGURES

TABLES

ACKNOWLEDGEMENTS

We would like to thank Mark Casson, Brian Goodall and Ian Gordon for their insightful comments and encouragement, and Abi Gillett, Emma Robinson and Carol Wilmshurst for compiling the figures and typing much of the text.

1

THE SCOPE AND LIMITATIONS OF ECONOMIC ANALYSIS

INTRODUCTION

The aim of this book is to demonstrate that tourism is susceptible to economic analysis to a much greater extent than is suggested by current literature by using economic concepts to provide a review and critical evaluation of the literature on a wide range of aspects of the economics of tourism and to indicate useful directions for future research. Thus, the book was not conceived as simply a text aiming to introduce basic economic concepts and theories to non-specialists. It attempts to go further by demonstrating the subject's ability to strengthen the theoretical foundations of more descriptive, diffuse and pragmatic research approaches to tourism and by showing economics' potential to explain and predict tourism phenomena. By examining tourism, using methodologies from mainstream and alternative schools of economic analysis, it also contributes new material about an activity of major and increasing economic importance, hitherto neglected in economics literature. In effect, the rationale of the book is to introduce more advanced applications of economic theory to tourism and so go beyond an introductory exposition of principles. For example, tourists' expenditure decisions, the structure of tourism markets and nature of decision-making within them, cross-country linkages between tourism firms, the extent and effects of foreign currency generation by tourism, the contribution of environmental resources and their relevance to policies for sustainable tourism have not been fully investigated and would benefit from further economic analysis.

The application of economic concepts, theories and methods raises fundamental issues concerning both the characteristics of tourism which distinguish it from other commodities and the core elements of economics as a discipline. Tourism is distinct, even as a service activity, in that it is usually purchased without inspection, consisting of a range of goods and services which are consumed in sequence, including transportation, accommodation and natural resources. Because natural and human-made resources are a substantial proportion of total inputs, it comprises a set of industries and markets involving significant non-priced features, such as the positive benefits which are associated with attractive and freely available environments and the negative effects of pollution (positive and negative externalities). It is, thus, the unusual composite nature of tourism, frequently taking place in different countries, which requires specific analysis.

Economics is characterized by alternative schools of thought and the predilection for economists to disagree is almost axiomatic. However, far from undermining the relevance and importance of the subject to tourism analysis, the controversies should be perceived as healthy as they signify a willingness to re-examine accepted tenets. Thus another aim of the book is to bring to the attention of the reader a range of theories and approaches which can be used to examine issues in tourism which are of a substantive as well as conceptual, theoretical and methodological nature. To an extent, the debates within economics are a reflection of its development which, in turn, reflects problems which arise within changing economies. Tourism is an activity which is subject to rapid change as underlying conditions alter so that many of the issues which economists acknowledge and address are also issues in the analysis of tourism.

A long-running debate has occurred within economics because the system of thought and analytical method of much of mainstream thinking has emphasized the attainment of equilibrium outcomes. This feature of the subject has been criticized by those who consider that the associated restrictive assumptions have narrowed the accepted orthodoxy, thus divorcing it from real world issues and problems. For instance conventional consumer behaviour concepts have been challenged by economic psychology and traditional theories of the firm and markets called in question by alternative views, such as those propounded by the Austrian school, which consider dynamic and disequilibrium circumstances. The range of different approaches in the discipline (reviewed by Greenaway et al., 1996) suggests that a pluralistic attitude is required, with cross-fertilization of concepts, theories and methods within and from outside the subject.

The analysis of tourism provides an opportunity for applying economic concepts and theories using both conventional and alternative methodological frameworks. In this sense, tourism is a useful vehicle for evaluating the appropriateness of alternative schools of thought in the discipline and for testing the robustness of economic concepts and methods. On the other hand, economic analysis can also contribute to greater understanding of tourism since, as pointed out by those who have long worked in the field, a number of aspects of the subject suffer from a weak theoretical framework through lack of appropriate research (Sessa, 1984). Similarly, concern has been expressed that much writing on the subject has no firm sense of direction and is methodologically unsophisticated (D.G. Pearce and Butler, 1993), which can have detrimental consequences for management and development policies.

Although this book recognizes the need for rigour in tourism analysis and uses economics as a means for achieving this, it opposes the notion that tourism should be treated entirely as an economic activity. There is a measure of agreement with D.G. Pearce and Butler (1993) when they state that each discipline which has studied tourism brings with it all the conceptual and methodological baggage of the subject. Such an approach would limit the scope and breadth of the analysis. Economists, as well as researchers from other disciplines, need to accept that, given the nature of tourism, the analysis of some aspects requires a multidisciplinary approach which acknowledges its political, physical and social contexts. Thus, it may be necessary to combine the findings of investigations by those from specialist disciplines and forge a more widely embracing theory. It is also reasonable to assert that methodologies frequently require modification to suit the circumstances under consideration. Tourism possesses characteristics which are peculiar to it, raising specific analytical and policy questions which force the investigator to reappraise the tools of the discipline. For example, decisions to purchase holidays are often made with friends and family so that mainstream consumer demand theory, which is based on individual decision-making, must also take account of individuals' and groups' social contexts. On the supply side, tourism relies on natural as well as human-made resources. This raises questions relating to the valuation of environmental resources which are freely available and potentially subject to degradation through excessive use. An associated problem area for economics is the increasingly important status accorded to sustainability and the possible trade-off between sustainable development and economic growth.

An important objective has been to make the book as accessible as possible, both to an economic audience and to those to whom economic concepts and methods are unfamiliar. To this end, theories and analytical approaches have been discussed without excessive use of technical terms and methods in economics or numerous descriptive statistics. The topics covered are necessarily selective. Two significant factors in the selection are, first, the desirability of including the main areas which are relevant to a comprehensive analysis of tourism, and second, the perception of key issues in tourism. Thus, the topics covered include demand, supply, market structure, pricing, output, growth and the environment, in a domestic and international context. The emphasis is on examining the role of quantifiable variables in tourism phenomena as this is where the contribution of economics to tourism research is likely to be most significant. It is important to note that the analysis of tourism within the book proceeds from the simple to the more complex in that economic concepts are initially used to examine tourism as though it were a non-composite commodity, similar to other goods, purchased by a relatively homogeneous set of consumers. Further complexities are gradually introduced throughout the book as attention is paid to the different components of tourism, initially in a closed economy and subsequently in an international context, and the issues of non-priced environmental resources and related market failures are finally considered. In the rest of this chapter the development of tourism analysis is traced and related to recent research from an economic standpoint. This acts as a starting point for the remainder of the book.

THE DEVELOPMENT OF TOURISM ANALYSIS

With very few exceptions, notably Gray (1966), there was little analysis of tourism until the 1970s and even then much research still tended to be uncritically descriptive, employing inadequately defined goals and techniques (D.G. Pearce and Butler, 1993). A considerable number of disciplines were and still are engaged in tourism analysis, the principal ones being anthropology, ecology, economics, geography, politics, social psychology and sociology. The main themes which have been studied are the cultural, economic, environmental and social effects of tourism, travel patterns and modes between origins and destinations, tourism's relationship with economic development, tourists' motivations and behaviour, forecasting trends and practical aspects of planning, management and marketing. Reference to economic studies is put on one side for the moment as discussion of these is undertaken later. No attempt is made to conduct a comprehensive review of the tourism literature as this has been carried out elsewhere (for example, Sheldon, 1990; Eadington and Redman, 1991; Sinclair, 1991a; Van Doren et al., 1994). However, it is useful to consider briefly the nature of the contributions to the development of the subject which have characterized the literature.

Most journal papers deal with specific aspects of tourism, very often case studies, without relating the topic to a particular analytical framework or theory. Indeed the tourism journal literature has been curiously uncontentious. A notable exception is the exchange that occurred between Gray and Sessa in the early 1980s (Gray, 1982, 1984; Sessa, 1984). Significantly, however, this was not directly about tourism analysis but the contribution of economics to it. Although it is in the journal literature that one can frequently discern the progression in the analysis of a subject, significant discussion of tourism issues and some analytical development has also occurred in books, any analysis being based on the concepts and methods of a particular discipline (Tisdell et al., 1988; D.G. Pearce, 1989). For example, the analysis of travel patterns and modes has been dominated by geographical analytical frameworks, while the study of demand outside economics tends to be underpinned by psychological or social psychological methods (D.G. Pearce and Butler, 1993). What emerges, therefore, is a number of studies which make a contribution to the literature by addressing a specific issue or attempting to develop a more analytical framework.

Several texts have provided descriptions of the structure and operation of different sectors of tourism. A well-known example is the volume by Burkart and Medlik (1989) and the 1980s saw a surge in such contributions to the literature, representative texts being those of Cleverdon and Edwards (1982), Hodgson (1987), Holloway (1994) and Lundberg (1989). The publications were not strongly analytical and were generally directed at the non-specialist reader. The focus on analysis became sharper in the late 1980s when texts such as those by McIntosh and Goeldner (1990), Ritchie and Goeldner (1987) and S.L.J. Smith (1989) appeared. A trend in the 1990s has been the collection of papers or readings in edited texts, some of which are essentially reviews of the state of the art. Although ostensibly aimed at practitioners, the book by Witt and Moutinho (1994) contains many chapters which act as a useful introduction to aspects of tourism which are of current concern. Other publications deal with both analysis and evidence. Typical examples are those by Sinclair and Stabler (1991), Johnson and Thomas (1992a, 1992b) and Seaton (1994). Cooper (1989, 1990, 1991) and Cooper and Lockwood (1992) provide useful reviews of several tourism topics from the perspectives of different disciplines. There is considerable variation in the rigour of contributions to such collections, particularly where there is a strong applied content reflecting the interests of practitioners.

Johnson and Thomas (1992a) focus on demand, dealing with tourist typologies, motivation, determinants of choice of activities and demand and forecasting models, including some quantitative perspectives. The contributions to Johnson and Thomas (1992b) are more applied and policy-oriented and mainly consider the impact of tourism on the economy and the environment. The topics covered are disparate and are not generally concerned with the analytical development of tourism. The collection edited by Seaton (1994) resembles those by Johnson and Thomas in that the nature of the papers varies greatly. Nevertheless, an indication of what are considered key topics by those undertaking research in tourism does emerge. The more general topics discussed include development, business, marketing and research, human resource management, the environment and tourism and society. These are extremely broad categories from which there are notable omissions. For example, most of the chapters on development omit an analytical base, the majority of papers being concerned with specific case studies. Similarly the section dealing with tourism businesses considers such aspects as gambling, museums and heritage sites, charter travel, airline pricing and information technology where the issues appear to concern trends and their policy implications. Consumer behaviour trends and business strategies emerge as themes in the discussion of marketing and research. The sections on human resource management and the environment concentrate on practical matters concerning, respectively, training and the means by which firms might maintain their viability in the face of the increasing problems of environmental degradation caused by tourism. Examination of tourism and society is almost entirely devoted to case studies of visitor–host interrelationships and socio-cultural change.

The volumes edited by Cooper are more cohesive to the extent that specific themes were identified at the outset and many of the chapters relate to them. The series has been particularly concerned with practical problems in recreation and hospitality management but has considered a wide range of additional topics including holiday choice and consumer behaviour, demand forecasting, urban tourism, sport, development, marketing, competitiveness, environmental issues, impact analysis and tourism statistics. Academic contributors have employed the theoretical and methodological stance of their disciplines. On the other hand contributors coming from the private sector, for instance, the hospitality, facilities provision and marketing sectors, tend to discuss practical problems requiring solutions of use to management. The general conclusion to be drawn is that the edited volumes indicate the concerns of academic researchers and practitioners but do not provide in-depth analyses of conceptual, theoretical and methodological issues.

Texts of a more applied character, mainly on management and marketing, are a dominant feature of the tourism literature. A fairly early example by Wahab (1975) is representative of the views then held as to the scope and nature of issues in tourism businesses. Typical examples tracing the development of applied tourism research are Hawkins et al. (1980), Foster (1985), Ritchie and Goeldner (1987), Laws (1991) and Witt and Moutinho (1994). Given that many students are preparing for careers in tourism, it is not surprising that numerous texts relate to the day-to-day concerns of businesses rather than the establishment of a theoretical base. Marketing is a case where many of the authors have worked in the private sector; well-known texts are those by Holloway and Plant (1988), Jefferson and Lickorish (1988) and Middleton (1988). Examples of more specific focuses on marketing are texts by Buttle (1988) regarding hospitality, Goodall and Ashworth (1988) covering historic cities, seaside resorts, travel agents and tour operators, Jefferson (1990) concerning tourist offices, Pattison (1992) with respect to the coach industry and S. Shaw (1987), who considered air transport.

Volumes which are directed at a more academic audience and have a stronger theoretical foundation have concentrated on spatial factors, the impact of tourism economically, environmentally, socially and politically, tourist behaviour, planning and development. A relatively early contribution to the spatial literature which considered both the pattern of development and origin-destination linkages was Robinson (1976). Later texts by Lozato (1985), Mill and Morrison (1985), D.G. Pearce (1987, 1989) and S.L.J. Smith (1983) continued these two related themes. Attempts to model travel are well reviewed by D.G. Pearce (1987), who identifies chronologically C.K. Campbell (1967), Yokeno (1968), Plog (1973), Miossec (1976), Britton (1980), Thurot (1980) and Lundgren (1982) as significant contributors to the development of tourism models. Page (1994) specifically considered transport for tourism, and the effects of transport improvements on tourism, notably the Channel Tunnel, were discussed by Page and Sinclair (1992a, 1992b), Page (1993), Sinclair and Page (1993) and Vickerman (1993).

The impact of tourism is an area of analysis which has attracted considerable interest. Deferring consideration of texts by economists until later, some notable contributions to the field are Mathieson and Wall (1982) who analysed the economic, physical and social effects of tourism, Jafari (1987), Murphy (1985a) and S.L.J. Smith (1989) who investigated social factors, Cohen (1978) and Shelby and Heberlein (1984) who concentrated respectively on the physical environment and carrying capacity and the Organization for Economic Co-operation and Development (OECD, 1981a, 1981b) concerning the environment. D.G. Pearce (1989) provides a useful review of the many forms of impact arising from tourism.

The literature on tourism and the environment and its many facets, such as alternative/cultural/eco/green tourism and sustainability, has burgeoned since the mid-1980s, paralleling the increasing awareness of problems arising from economic growth and, with it, the expansion of tourism. Consideration of this issue has broadly gone in two directions. The first is tourist behaviour and its impact on the environment, with noteworthy contributions to analysis by Krippendorf (1987), Elkington and Hailes (1992), Cater (1993) and Wheeller (1994). The second is related to initiatives aimed at attaining sustainable development and focuses on tourism firms. Much attention has been paid to voluntary actions by businesses and much of the literature emanates from or is directed at practitioners, for example, Beioley (1995), Dingle (1995), Eber (1992), Hill (1992), Institute of Business Ethics (IBE, 1994), International Chamber of Commerce (ICC, 1991), World Travel and Tourism Council (WTTC, 1994), so that the emphasis is on prescriptions for management and practice. The academic literature is more rigorous and has endeavoured to derive principles, as well as to indicate feasible policies for achieving sustainability, for instance, C. Smith and Jenner (1989), Cater (1993), Wight (1993), Cater and Lowman (1994), Goodall and Stabler (1994), Murphy (1985b) and Hunter and Green (1995).

Tourist behaviour, in terms of both the determinants of demand in generating areas and activity in destinations, is another focus of study. The non-economic literature tends to concentrate on the motivation to take a holiday and the purpose of the trip. A substantial contribution to the literature has been made by sociologists, psychologists and social-psychologists, as well as geographers. An early and seminal psychological contribution to the analysis of tourist motivation and behaviour is that of Grinstein (1955). Gray (1970) introduced the terms 'wanderlust' and 'sunlust' as categories of tourists and purposes for taking holidays. Plog (1973, 1987) was also prominent in deriving motivational models of demand. Crompton (1979) developed the notion of motivations in an empirical study which later writers (for example, Dann, 1981; Graburn, 1983; Iso-Ahola, 1982; Leiper, 1984) have investigated. With respect to behaviour, there are representative studies by P.L. Pearce (1982) and Pizam and Calantone (1987). A useful review of motivations and behaviour is provided by D.G. Pearce and Butler (1993).

Tourism development, planning and policy-making has not been extensively investigated. A relatively early contribution was by de Kadt (1979) and others include Lea (1988), Williams and Shaw (1988), Page and Sinclair (1989), Dieke (1993a, 1993b), Sinclair *et al.* (1994) and D. Harrison (1995). D.G. Pearce (1989) contains a good overview of the subject. Interest in tourism planning is a recent phenomenon, possibly because of not only growing recognition of the need for it to be more systematic in the face of greater competition between destinations, but also recognition that the growth of tourism may need to be constrained because of its environmental and socio-cultural impact. Examples of literature on tourism planning are Gunn (1988), Getz (1986, 1992) and Inskeep (1991) and there has also been some discussion of regional tourism co-operation, for example, Teye (1988) and Dieke (1995).

Examination of the tourism literature reveals that the more rigorous development of research, with a concomitant extension of aspects of the subject studied, is under way. However, it is evident that more in-depth analysis of tourism is required and further examination of particular issues is necessary. Steps towards these objectives have been taken by G. Shaw and Williams (1994) who expressly set out to consider critical issues in tourism. They argue that, to date, the tourism literature has developed in a compartmentalized fashion and that points of contact between disciplines are uncommon, recognizing that many aspects of tourism have been neglected by economists. Economic issues in tourism are seen as arising from trends towards the increasing commoditization and privatization of leisure. While Shaw and Williams acknowledge that tourism has the capability of generating export earnings and facilitating economic growth and diversification, they consider that it has caused acute environmental problems and initiated the reconstruction, economically, physically and socially, of localities to serve tourists and the private sector. Tourism also involves the social construction of artefacts and activities considered attractive by visitors, with concomitant changes in lifestyles and social differentiation. The impact of tourists on destinations brings tourism firmly within the ambit of public sector economics with implications for taxation of firms and individuals and expenditure on infrastructure provision. They conclude that tourism has many features which characterize other economic sectors, for example product cycles, changes in labour processes, increasing concentration in tourism supply and the integration of firms across national borders.

In speculating on future trends, with possible research implications, Shaw and Williams suggest that firms' actions will be largely driven by consumers with increasing preferences for cultural and green tourism. They see a polarized market with the latter's tourism consumption contrasting strongly with that of the mass-market, bringing about the restructuring of tourism places with far-reaching capital, labour market and public policy effects. Shaw and Williams underline the view of earlier researchers such as Greenwood (1976) and Burkart and Medlik (1989) that tourism issues should not be addressed in isolation but must be set in the wider context of those pertinent to the

8

social sciences as a whole. A number of the issues raised are relevant in evaluating the economics literature on tourism, to which attention is now turned.

ECONOMIC ANALYSIS OF TOURISM

The literature on the economic analysis of tourism is curiously unbalanced in that numerous studies have been undertaken of some topics, notably demand, forecasting and multiplier studies of the impact of tourism, while little attention has been paid to others. Demand and expenditure on tourism have been investigated for a wide variety of international origins and destinations. For example, Gray (1966) considered international travel in the US and Canada, Barry and O'Hagan (1971) studied tourist expenditure in Ireland, O'Hagan and Harrison (1984) conducted a study of UK expenditure in Europe, Bechdolt (1973) and Witt (1980) modelled demand for a range of foreign holidays and Pack et al. (1995) examined the changing spatial concentration and dispersal of international tourism demand in the UK. The investigation of demand has generally involved the estimation of the relative importance of particular variables which determine the level and pattern of holiday expenditure, such as income, relative prices, exchange rates and transport costs. The many studies of tourism demand in different countries and time periods are reviewed by Archer (1976), Johnson and Ashworth (1990), Sheldon (1990) and Sinclair (1991a) while Witt and Martin (1989) examined alternative approaches to tourism demand forecasting. None the less, most of the studies of tourism demand which have been undertaken to date lack an explicit theoretical underpinning.

In contrast to the literature on demand, supply is a significant lacuna in economic studies of tourism. A range of accounts of the structure of tourism production have been provided and specific aspects of tourism supply have been investigated, for example concentration (McVey, 1986; Go, 1988, 1989), integration (J. Randall, 1986; Bote Gómez et al., 1991; Bote Gómez and Sinclair, 1991), and the market structure of particular sectors, notably tour operations (Sheldon, 1986, 1994; Fitch, 1987; N. Evans and Stabler, 1995). Figuerola (1991) examined the operation of the tourism firm from a microeconomic perspective. However, as yet there is no coherent theory of tourism supply and virtually no quantitative research on its determinants, its responsiveness to changes in them or on the complementarity or substitutability of the capital, labour and environmental inputs used in production. Accordingly, investigations have gone little further than description. A related area in which analysis is sketchy concerns the determination of prices, although hedonic pricing models have attempted to take some account of the complexity of tourism, as a composite product, in the price competitiveness of package holidays (Sinclair et al., 1990; Clewer et al., 1992).

The impact of tourism on income and, in some cases, employment has been examined for many tourist destinations. Archer (for example, 1973,

1989), Sadler *et al.* (1973) and Sinclair and Sutcliffe (1978, 1988a, 1988b, 1989a) made analytical as well as applied contributions in the area of multiplier modelling and innovative studies include those by Henderson and Cousins (1975), Varley (1978) and Johnson and Thomas (1990). Relatively little economic analysis of the employment effects of tourism has been undertaken, the studies by Vaughan and Long (1982), Goodall (1987) and Johnson and Thomas (1990) being notable exceptions. Official organizations concerned with economic development, such as the United Nations Development Programme (UNDP), have been concerned with tourism's ability to generate income and employment in developing countries and have incorporated the findings from a number of studies which have used the multiplier model into their assessment of the appropriate policy recommendations for the areas under consideration. A theoretical general equilibrium model of the effects of tourism was developed by Copeland (1991) and Adams and Parmenter (1995) subsequently estimated a computable general equilibrium (CGE) model to show the structural repercussions of tourism expansion. Zhou *et al.* (1997) estimated the economic impact of tourism on Hawaii using CGE and input–output analysis.

International aspects of tourism, such as the role of multinational corporations and other forms of cross-border linkages between firms (Dunning and McQueen, 1982a; Sinclair *et al.*, 1992), the terms of trade (Curry, 1982), instability of foreign currency earnings from tourism (Sinclair and Tsegaye, 1990) and the relationship between tourism and economic development (exemplified by Bryden, 1973; Varley, 1978; Britton, 1982; Aislabie, 1988a; Bachmann, 1988; Theuns, 1991; D. Harrison, 1992; Oppermann, 1993; Dieke, 1995; Sinclair, 1997a) have received some attention. A main focus of research has been tourism's contribution to foreign currency earnings and the balance of payments (UNCTAD, 1973; English, 1986; G. Lee, 1987) and the need for more accurate statistics and, hence, more accurate balance of payments accounts (Grünthal, 1960; Instituto Español de Turismo, 1980, 1983; White and Walker, 1982; Baretje, 1982, 1987, 1988). With a few exceptions (Gray, 1970, 1982; Socher, 1986; Vellas, 1989) little attention has been paid to explaining inter-country tourism flows using international trade theory.

Other aspects of tourism in which economics has hardly begun to be involved concern the environment, fiscal policy and regulation. There is a growing literature regarding tourism and the environment but most studies lack a strong economic base, concentrating on cultural, physical and social aspects or on business initiatives regarding sustainability. Economists have tended to direct their attention at leisure and recreation, emanating from pioneering work by Clawson (Clawson and Knetsch, 1966), especially the problem of valuation of non-priced goods and services. Useful overviews are contained in Walsh (1986) and Hanley and Spash (1993). The limited number of studies focusing explicitly on tourism have concentrated on particular topics; for example, Wanhill (1980) examined the feasibility of charging to reduce congestion at popular attractions and a little later evaluated the resource costs of tourism (Wanhill, 1982).

Developments in environmental economics with respect to pollution, resources depletion, degradation of the natural and built environment and sustainability have generally made only passing reference to service sectors such as tourism (D.W. Pearce *et al.*, 1989; Pearce and Turner, 1990; Turner *et al.*, 1994). Attention has focused on evaluating the benefits of natural and human-made resources resulting in a vast literature on the most appropriate methods, such as the contingent valuation, hedonic pricing and travel cost models. A standard text on the contingent valuation method is R.C. Mitchell and Carson (1989) while hedonic pricing has been applied to historic buildings, heritage sites and amenity resources, for example by Willis and Garrod (1993a), and to package holidays by Sinclair *et al.* (1990). The travel cost method has been the most widely used approach to evaluating leisure amenities and there are numerous examples over a thirty-year period (for example, Clawson and Knetsch, 1966; Burt and Brewer, 1974; Cheshire and Stabler, 1976; McConnell, 1985; Willis and Benson, 1988; Willis and Garrod, 1991a; Bockstael *et al.*, 1991; Hanley and Ruffell, 1992). A rare instance of the application of the contingent valuation and travel cost methods to tourism is that by Brown and Henry (1989). Sinclair (1992a) has provided an overview of economics-related literature on tourism, development and the environment and Brooks *et al.* (1995) and Stabler (1995a) have reviewed the main economic methods in a study concerned with urban conservation and its role, together with that of tourism, in regenerating urban economies.

Public sector economics has virtually ignored the impact of tourism on national and local economies and the potential for national and local public finance policy to mitigate its detrimental effects on the environment or to fund the required infrastructure and services. There are, however, some examples of research on feasible forms of taxation. There have been contributions on accommodation tax (Mak and Nishimura, 1979), types of tax (Fish, 1982; Bird, 1992), equity effects (Hughes, 1981; Weston, 1983) and hotel room occupancy, entertainment and excise (sales) tax (Fujii *et al.*, 1985). The idea of differential taxes intra-nationally has been investigated (Jeffrey and Hubbard, 1988) as have applications at the national level (Archer, 1977b). While taxes increase the costs of tourism, subsidies and grants can be employed to stimulate supply; in this context Wanhill (1986) has examined the role of capital grants, ongoing operational subsidies and tax relief. Regulation is also a means of altering the repercussions of tourism. For example, adverse effects of tourism on the environment can be significant and research has been conducted in the environmental field regarding regulation among other means of dealing with them. However, studies have concentrated on manufacturing and the effectiveness of regulation in comparison with price-based instruments, for example Tietenberg (1988), Turner (1988), Opschoor and Vos (1989), Opschoor and Pearce (1991),

Opschoor and Turner (1993), O'Riordan (1992), Stavins (1988) and Turner *et al.* (1994).

During the early 1980s, Gray (1982, 1984) and Sessa (1983, 1984) reviewed the contribution of economics to the analysis of tourism and identified key limitations of the methodological approach as being the need for periodic critical examination of economic assumptions and hypotheses, the weakness of its theory in a spatial context and the need for research to be more systematic. Gray also argued that tourism's extensive use of natural resources and public utilities, as well as the temporal and spatial variations in demand, raised questions as to the efficacy of economic methods. A further overview of themes in tourism research was undertaken by Aislabie (1988b), who concentrated mainly on tourism and economic development, its impact on communities and the environment and the related implications for public policy. Aislabie identified the main shortcoming of tourism research as lack of depth and also argued that there was an insufficient nexus of tested and agreed propositions to allow generalizations to be made. He considered a number of topics as central to economic contributions to tourism analysis, namely organization of production, matching demand and supply through markets, price formation, impact methodology, forecasting and tourism in developing countries. However, he alighted on specific aspects which impede understanding of tourism, such as the difficulties created by the lack of clear definition compounded by inaccurate statistics, the complexities of the characteristics of the product, unstable demand, the excess capacity/peak load problem, pricing and marketing, in the context of incomplete information. Aislabie recognized that tourism is characterized by a market structure which is peculiar to it and which poses acute analytical difficulties, pointing out that although these topics appear to be amenable to conventional analysis, research has failed to make significant headway in the field.

Little has changed in the early to mid-1990s. Although textbooks on the economics of tourism have been published, for example Bull (1991), Cooper *et al.* (1993), Burns and Holden (1995), Lundberg *et al.* (1995) and Tribe (1995), they do not claim to be in-depth applications of the conventional principles of economics to tourism. While the volumes are useful as basic texts, the relevance and application of the subject are not shown fully, being confined largely to simple demand and supply analysis, outlines of market structures and elements of forecasting and pricing such as price and income elasticities. It is against this background, where economic analysis has been undertaken only of specific topics in tourism, that this book was conceived.

SCOPE OF THIS BOOK

This book reflects what are perceived as core issues, both in economics and tourism. The approach is to consider tourism demand and supply using alternative analytical frameworks before examining tourism in an international and environmental context. Throughout the book the objective is to introduce the reader to recent developments in economic analysis and to indicate their relevance to tourism issues through illustrative examples. The analysis initially follows the existing tourism literature by considering tourism as an aggregate commodity, without examining its components. Thus, Chapters 2 and 3 consider theories and applied studies of the demand for tourism, disaggregated only by origin and destination nationality. In Chapter 2, economic concepts are used to explain the demand for tourism and the effects of changes in income and relative prices, as well as the timing of tourism demand. The social context of tourist decision-making is also investigated and developments outside mainstream theory, such as research into the motivation to take a holiday by economic psychology, are examined. Models which have been used to estimate the responsiveness of demand to changes in income, relative prices, exchange rates and one-off events such as EXPO are explained and critically examined in Chapter 3 and their relevance to methods for forecasting demand is indicated. Suggestions for further research, particularly at the microeconomic level, are made.

Chapters 4 and 5 investigate tourism supply, recognizing its peculiarities as a service sector activity, an aspect of supply which tends to be overlooked in economics. The discussion introduces additional complexities into the analysis by ceasing to regard tourism as an aggregate commodity and, instead, examining such components as accommodation and transport. Initially in Chapter 4 an exposition of the conventional economic analysis of firms and markets is provided as it offers significant insights into the competitive structure of tourism supply. The major sectors of accommodation, the intermediaries (travel agency and tour operators) and transport are examined and a number of features which characterize tourism supply are identified. It is argued that these features and the complex structure of tourism markets require explanations which are additional to those given by conventional analysis. Therefore, in Chapter 5, other approaches to describing and understanding the structure and operation of tourism supply are considered, largely within the context of an industrial economics paradigm. Particular reference is made to concentration and entry/exit conditions and the tour operations sector is examined as an illustrative case. Recent analytical developments are introduced and related to tourism markets to explain firms' strategies and interrelationships. Special attention is paid to game theory to reflect the dynamic and often disequilibrium nature of tourism markets. These are characteristics of tourism supply which severely test the relevance of conventional analysis, especially regarding the complex structure of tourism markets.

In Chapter 6, the discussion is broadened to consider tourism demand and supply in the international arena. The first part of the chapter considers explanations of international tourism. Theories of comparative advantage are outlined, providing insights into tourism in competitive market conditions. The relevance of the demand for ranges of qualities of tourism products in imperfectly competitive markets, economies of scale and scope, tourism product differentiation, competition via research and development and the international interdependence of tourism firms is then discussed. Explanations of the increasing dominance of large multinational firms, such as international hotel chains, are offered. The second part of the chapter examines some of the economic effects of international tourism, including income and employment generation, foreign currency provision and growth.

Chapters 7 and 8 consider environmental issues in tourism, especially in the light of consumer concerns and the stated commitment to the principles of sustainable development. The economic analysis of the previous chapters is extended to take account of the public good (freely available and accessible) nature of many tourism resources and of market failures in such forms as environmental pollution. An overview of the principal environmental issues is given from an economic standpoint in Chapter 7, where the development of both the general and tourism sustainability concept is traced. An analytical framework for appraising resource use and the impact of tourism activity is proposed and key environmental issues are identified. The elements of conservation economics, which underpin the objectives of sustainability, are then explained and related to the use of productive, energy, human-made and natural resources in tourism. In Chapter 8, economic methods for measuring the non-market environmental benefits and costs of tourism and the instruments for achieving sustainability are explained and evaluated. The economic concept of market failure is defined and discussed as a justification for intervening in tourism markets in order to secure resource conservation and environmental protection. The effects of the instruments and the implications for policy are outlined. The final chapter in the book is concerned with the implications, for research, of the analysis of the range of issues covered in the previous chapters.

Owing to the usual constraints of time and space, a number of important topics have received only cursory attention. These include the spatial aspects of tourism, labour markets, migration and population displacement, tourism and economic development and the distributional effects of tourism. The practical aspects of tourism with respect to marketing and management are not specifically considered. Reference is made to the fiscal and regulatory functions of the public sector and some planning implications are outlined but not examined in depth. Although its importance is acknowledged, little mention is made of business tourism. All of these areas provide ample scope for further research.

2

THE MICROFOUNDATIONS OF TOURISM DEMAND

INTRODUCTION

The relative and absolute importance of tourism in people's expenditure budgets has risen dramatically, with consequences not only for the welfare of tourists themselves but also for the residents of the areas they visit. The large numbers of tourists and the scale of their expenditure has considerable effects on the income, employment, government revenue, balance of payments, environments and culture of destination areas. A fall in demand can bring about decreases in living standards and rises in unemployment, while increased demand can result in higher employment, income, output and/or inflation and may threaten environmental quality and sustainability. Furthermore, tourism firms are confronted by changing revenue and profits and governments experience changing tax revenue and expenditure. Thus, tourism demand affects all sectors of an economy – individuals and households, private businesses and the public sector.

The significant level and repercussions of tourism demand provide a strong case for better understanding of the nature of tourists' decision-making process. A further reason for doing so is that policies which are formulated in relation to tourism demand ultimately depend upon the relevance of the theories which have been used to explain and estimate it. The incorporation of an inappropriate theoretical framework in empirical studies of demand can result in incorrect specification of the equations which are used to estimate tourism demand and biased measures of the responsiveness of demand to changes in its determinants. Any policy measures which are based on such measures are also likely to be misguided.

Accordingly, this chapter will examine the economic theories which underlie tourist decision-making at the microeconomic level while Chapter 3 will concentrate on explaining and evaluating the models which have been used to estimate and forecast demand. The discussion in the two chapters differs from past economic analyses of tourism demand by providing a theoretical framework which can be used to evaluate the many empirical studies which have been undertaken. It should also provide the basis for the formulation of appropriate models for estimating tourism demand and, hence, for more accurate results and more appropriate policy implications. The discussion will introduce recent developments in the economic literature relating to consumer behaviour, particularly the microeconomic basis of more aggregate relationships at the national level and intertemporal preferences.

The emphasis in the theoretical analysis in this chapter is on variables which can be measured quantitatively, including the effective demand for tourism which is the amount that consumers are willing and able to spend rather than the notional demand which they would like to exercise but which is not backed by the ability to pay. This does not imply that non-measurable, qualitative variables are unimportant. Indeed, research in economic psychology takes some cognitive variables into account and mainstream economics acknowledges the important role which expectations can play. However, measurable 'material' variables are the main focus of attention since this is the area where economics has most to contribute. Thus, the approach which is taken in the two chapters complements the non-economic analyses of tourism demand which have been undertaken to date.

The first part of this chapter will examine the economic theory which explains tourism demand. Initially, the relationship between employment, income, demand for consumer goods and unpaid time is investigated. This provides an explanation of a person's demand for all consumer goods and services, including tourism. The choice of how much tourism to purchase relative to other goods and services is then discussed and the analysis is then extended to consider choices between different types of tourism, including the cases where one type of tourism complements or substitutes other types. The effects of changes in income and relative prices on tourism demand are examined and some determinants of the timing of spending are then explained. Although the theoretical explanation in the first part of the chapter may appear somewhat technical, it is crucial in providing a rigorous basis for applied work. Empirical studies which are undertaken without an explicit theoretical underpinning may produce biased results with misleading policy implications for the area concerned. The second part of the chapter departs from the earlier emphasis on the individual by considering the social context of tourism decision-making, showing how insights from other disciplines can contribute to the economic analysis of tourism. The relationship between the theory of tourism demand and empirical studies will be discussed in Chapter 3.

OPTIMAL CHOICE IN TOURISM DEMAND

Consumption, paid work and unpaid time

Both people's preferences and their expenditure budgets are key determinants of the demand for tourism. A person who is considering whether to spend a holiday away from home has an amount of money, or budget, which is available for expenditure on tourism and other goods and services. The size of the budget depends upon the number of hours that he or she spends in paid work per time period (labour supply), on the income per hour and on the rate of taxation on income which yields the disposable income available for purchasing goods and services. People trade-off paid work against unpaid time; some people prefer more income, resulting from more paid work, while others prefer to have more unpaid time for leisure or household activities and therefore spend less time in paid work. If they undertake more paid work and have less unpaid time, their level of income rises but leisure and household work are foregone; conversely, taking more leisure reduces income. There is, however, a tension as income is often required to undertake leisure pursuits so that the latter have an imputed 'price' or opportunity cost. Each combination of paid work and unpaid time provides a different amount of earnings, or budget, which may be spent on goods and services. The highest ratio of paid work to unpaid time usually provides the largest budget, corresponding to the largest potential consumption value, and vice versa.

The different combinations of consumption and unpaid time which a person may have are illustrated by the line CBU in Figure 2.1. The vertical axis measures the value of consumption and the horizontal axis measures increases in unpaid time when read from left to right (or increases in paid time when read from right to left). The point OC shows the maximum consumption which the person can achieve, resulting from spending the maximum possible time in paid work. Someone who is not in paid work has a

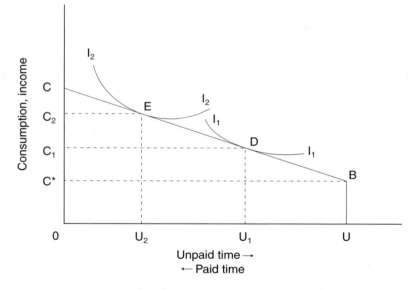

Figure 2.1 Consumption, paid work and unpaid time

17

combination of consumption and unpaid time shown by B, with OC* being the value of consumption which the person obtains while unemployed, for example, by unemployment benefits. Positions between C and B show intermediate combinations. The line CBU is known as the budget line, the slope of which indicates the rate of remuneration; so if, for example, the wage rate rises, it becomes steeper.

People also receive satisfaction, or utility, from consumer goods and from unpaid time. For example, a person may receive the same utility from a large amount of consumer goods and small amount of unpaid time as from a small amount of goods and high amount of unpaid time, or from intermediate combinations of the two. The different combinations of consumption and unpaid time which provide a given level of satisfaction are depicted by the curves I_1I_1 and I_2I_2 in Figure 2.1. The curves are known as indifference curves because the person is indifferent between alternative positions on a given curve, since he or she receives the same satisfaction from every choice of quantities of goods and services depicted by it. Curves further away from the origin of the graph would correspond to higher combinations of consumption and unpaid time, and hence to higher satisfaction, and vice versa.

Economists assume that people want the maximum satisfaction possible and obtain this by choosing the combination of paid work and unpaid time which provides them with their preferred, attainable combination of consumer goods and unpaid time. One person's preferred position might be at D in Figure 2.1, which is determined by the point of tangency between the indifference curve I_1I_1 (determined by the person's preferences) and the budget line and shows the person's optimal combination of consumption, OC_1, and unpaid time, OU_1 (with paid work time UU_1). An individual with different preferences would have an alternative combination of consumption and unpaid time. For example, the individual might obtain the same level of satisfaction from higher ratios of consumption and paid work to unpaid time, depicted by the curve I_2I_2. Thus, the individual's optimal position, with given preferences and income per hour, would be at E, corresponding to consumption of OC_2 and a lower value of unpaid time, OU_2 (and higher paid time UU_2). There are, of course, circumstances when a person is unable to choose his/her preferred combination of consumption and unpaid time and is constrained by structured hours of work to a specific sub-optimal position, for example because of having to work a thirty-eight-hour working week with no possibilities of part-time work or overtime. Moreover, he/she might be forced to take additional unpaid time because of a downturn in the economy or might even be made redundant with unpaid time OU and consumption OC*.

The amount which is available for spending on tourism and other goods is, therefore, the income or budget which results from the person's paid work (labour supply) and from his or her preferences between consumption (permitted by paid work) and unpaid time. Hence, consumption and labour supply are jointly determined and should be considered simultaneously. Changes

18

in remuneration for work bring about changes in people's consumption and unpaid time. For example, an increase in the wage rate or a decrease in income tax results in higher income and greater consumption and more, less or the same amount of unpaid time. This is because an increase in effective remuneration per hour encourages a person to substitute higher paid work and higher consumption for unpaid time, i.e. the substitution effect. Conversely, the person can use higher earnings from a given amount of paid work time to purchase more goods while simultaneously taking more unpaid time, the income effect. The net effect is that depending on personal prefer- ences, he/she can increase consumption while having more, less or the same unpaid time. The application of the concepts of substitution and income effects to tourism will be considered following discussion of the extent to which people demand tourism *vis-à-vis* other goods and services.

Demand for tourism relative to other goods and services

The demand for tourism depends upon the total budget which is available for spending (resulting from the person's labour supply or unemployment bene- fits, as discussed above) and on preferences for tourism relative to other goods and services. At one extreme, the person could allocate all of his/her budget to tourism and, at the other, none of it to tourism and all of it to other goods. Between these two extremes, a range of combinations of tourism and other goods are feasible. All possible combinations are given by the budget line, the slope of which indicates the relative prices of goods and services and which is depicted by TG in Figure 2.2. The point OT is the

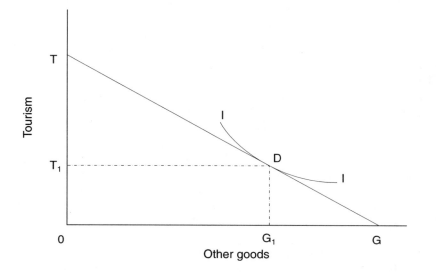

Figure 2.2 Consumption of tourism and other goods

amount of tourism which would be consumed if a person spends all of his/her budget on tourism and OG is the amount of other goods which would be consumed if there were no expenditure on tourism, with the line TG showing intermediate combinations. The amounts of tourism and of other goods which it is possible to consume depend upon the relative prices of tourism and other goods so that lower tourism prices would permit more tourism and vice versa. The effect of changes in relative prices, which are depicted by changes in the slope of the budget line, will be discussed below.

The combination of tourism and other goods which the person decides to purchase depends upon his/her preferences. Alternative combinations of tourism and other goods can provide the consumer with the same level of satisfaction so that, for example, low tourism consumption and high consumption of other goods provides the same satisfaction as high tourism consumption and low consumption of other goods, as illustrated by the indifference curve II in Figure 2.2. The person allocates his/her budget between tourism and other goods by choosing the combination which maximizes satisfaction. This is the point D, where the indifference curve is tangential to the budget line, resulting in OT_1 tourism and OG_1 consumption of other goods. An individual with a stronger preference for tourism would consume a combination to the left of point D, whereas someone keener on consuming other goods would have an indifference curve tangential to TG to the right of D.

People have to decide not only on their preferred combination of tourism relative to other goods, but also on their preferred combination of different types of tourism. For example, a tourist could spend all of his/her tourism budget on visits to friends and relatives or all of it on holidays in new locations abroad, or could choose some combination of the two. The optimal position again depends upon the person's budget and preferences and it is again assumed that the budget is allocated between different types of tourism so as to maximize satisfaction. The optimal combination of visits to friends and relatives and holidays abroad can be illustrated using a graph similar to Figure 2.2 but with the different types of tourism measured on the axes and is shown in Figure 2.3. In reality, of course, there may be more than two combinations; these could be shown mathematically but not diagrammatically.

In the case of different types of tourism, a person may choose a combination of tourism types. However, this is not the only outcome which may occur as one type of tourism may be a substitute for or complement to another. For example, some American tourists who travel to Europe regard destinations in different European countries as a complementary part of the tourist experience, rather than substitutes, so that, for example, London and Paris may be regarded as complements and fixed proportions of expenditure are allocated to each. This case is depicted in Figure 2.3, where the budget line T_PT_L shows how different combinations of tourism expenditure could be allocated to the two destinations, but the L-shaped indifference curve II shows that the person wishes to allocate set proportions of the budget to each.

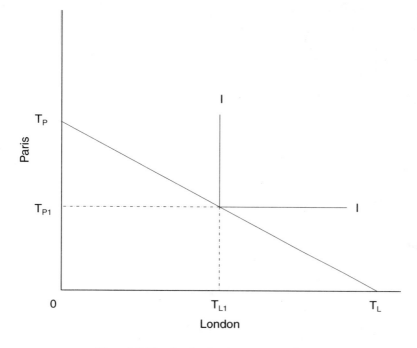

Figure 2.3 Tourist destinations as complements

The alternative case of tourist destinations as substitutes might apply to holidays in Sydney and New York, as illustrated in Figure 2.4. The budget line, $T_S T_{NY}$, indicating the relative prices of the two holiday destinations, again shows that different proportions of the budget might be allocated to tourism in each destination. However, the indifference curve $I_B I_B$ shows that person B regards the two destinations as substitutes and selects New York as the preferred destination. Another individual, C, also regards the two destinations as substitutes but has different preferences, illustrated by the indifference curve $I_C I_C$, and chooses Sydney rather than New York. Knowledge of the extent to which different types of tourism or tourist destinations are substitutable or complementary is particularly useful for tourism planning and marketing but this is an issue that has hardly begun to be explored in the tourism literature.

The effects of changes in income and prices on tourism demand

Economists posit that tourism demand is affected principally by income and prices and information about the extent to which changes in demand result

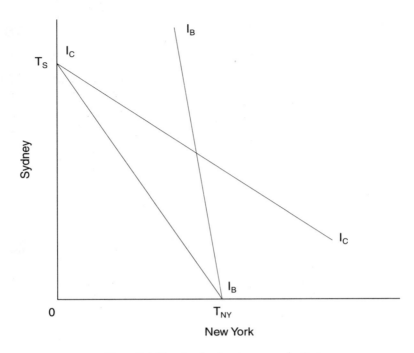

Figure 2.4 Tourist destinations as substitutes

from each of these variables is also important for both tourism suppliers and policy-makers. It is helpful, initially, to examine the effects of each of these variables separately. In the case of a rise in income with constant relative prices, the effect on most types of tourism and most tourist destinations is likely to be positive. Thus, an increase in income results in a rise in tourism purchases, similar to the effect of increasing income on the demand for most other goods and services; i.e. it is a *normal* good because demand for it is positively related to income. However, it is possible for a rise in income to bring about a fall in demand, such as for tourism in mass market destinations, implying that this form of tourism is an *inferior* good. This might be the case where a beach holiday in the Caribbean is substituted for one in the Costa Brava.

The two effects are illustrated in Figure 2.5. The vertical axis measures tourism and the horizontal axis measures other goods. The lines TG and T'G' are the budget lines before and after the rise in income respectively and are parallel because of the assumption of constant relative prices for tourism and other goods. Indifference curves are included to illustrate the person's preferences. If tourism is a normal good, preferences may be illustrated by the indifference curve I_2I_2 so that demand rises from OT_1 to OT_2 at E. If it

22

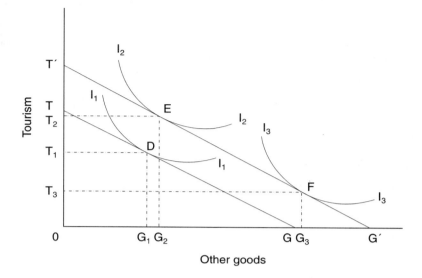

Figure 2.5 Effects of a rise in income on tourism consumption

is an inferior good, indicated by the indifference curve I_3I_3, an increase in income brings about a decrease in tourism from OT_1 to OT_3 at F. If demand is positively related to income and rises more than proportionately, the good is known as a *luxury* and if demand rises less than proportionately, it is known as a *necessity*. In terms of the concept of elasticity, the demand for a luxury good is said to be elastic with respect to changes in income while that for a necessity is inelastic.

The second case to be considered concerns the effect on tourism demand of a change in relative prices with income held constant. Demand and prices are usually negatively related, so that a fall in price is normally associated with a rise in demand and vice versa. The effect of a decrease in the price of tourism is depicted in Figure 2.6. Since tourism is now cheaper, the person's budget can now purchase a maximum of OT' tourism instead of OT, while the maximum amount of other goods which can be purchased remains constant at OG since their prices are assumed to remain constant. The combinations of tourism and other goods which can be purchased after the price decrease are given by the line $T'G$. The original and subsequent optimal combinations of tourism and other goods are respectively points D and E in Figure 2.6, so that a fall in the price of tourism results in an increase in demand and satisfaction as the person purchases OT_2 tourism and OG_2 other goods as opposed to OT_1 and OG_1 before the fall in price. It is also possible to consider the choice between two similar forms of tourism, where the price of one changes relative to the other. Thus, for example, a resident

23

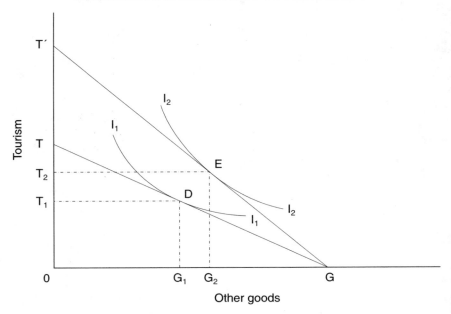

Figure 2.6 Effects of a fall in price on tourism consumption

of the UK may be contemplating one of two resort holidays in the Mediterranean, one in France and the other in Italy, but should the French franc appreciate in value against sterling while the lira remained unchanged, the Italian resort holiday would be chosen.

It is possible to depict both the income and price effects on tourism demand in the same diagram, as illustrated in Figure 2.7 which effectively combines Figures 2.5 and 2.6. Suppose, for example, that there is a change in the price of tourism so that tourism becomes cheaper relative to other goods and the person's budget line shifts from TG to T'G. The optimal point of consumption was originally at D. The effect of the change in relative prices, with income held constant, is demonstrated by drawing the broken line, PP, with the same slope as the new budget line T'G, and hence with the new relative prices, tangential to the original indifference curve I_1I_1. (Since the line is tangential to the original indifference curve, satisfaction and income are constant.) The effect of the change in relative prices is given by the move from D to S. This effect is known as the substitution effect, since the fall in the price of tourism has caused the person to substitute relatively cheaper tourism for other goods, so that tourism demand rises and the demand for other goods decreases.

The second effect is that of the change in real income; as tourism is now cheaper, the person is better off in real terms. He/she has the option of

24

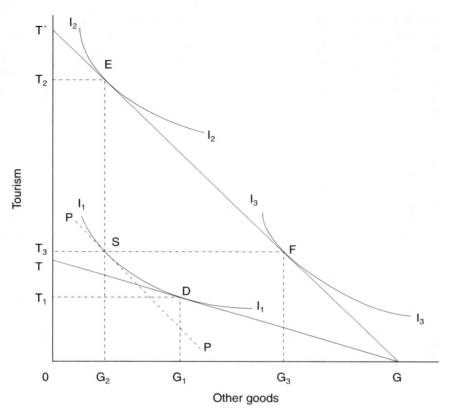

Figure 2.7 Effects of a fall in price and a rise in income on tourism consumption

spending all the increase in real income on tourism or all of it on other goods or some on each. If the person chose to spend all the increase on tourism, the income effect would be illustrated by a move from S to E in Figure 2.7 where OT_2 tourism and OG_2 other goods are purchased (since PP is parallel to T'G, the price ratio is held constant so that only the effect of the increase in income is taken into account). If all of the increase were spent on other goods, the income effect would be from S to F where OT_3 and OG_3 are purchased respectively. If he/she chose to demand more of both tourism and other goods, the optimal point would be somewhere between E and F. The net effect of the change in relative prices is the move from D to E, from D to F or from D to a point between E and F respectively. Hence, tourism demand rises while the demand for other goods falls, remains constant or rises, depending upon the person's preferences.

Before moving on from the above discussion, it is worth noting that the substitution effect resulting from the change in relative prices, as demonstrated in Figure 2.7, was defined with income (purchasing power) held constant. An alternative definition of the substitution effect takes account of the change in demand resulting from a change in relative prices, when utility rather than income is held constant. In the latter case, the demand for tourism is known as the compensated (rather than uncompensated) demand because the person is said to have been 'compensated' for the change in relative prices by allowing his/her level of satisfaction to remain the same.

Tourism demand over time

People's choices concerning the timing of consumption are known as intertemporal choice and have attracted increasing attention within economics (for example, Heckman, 1974; Obstfeld, 1990; Deaton, 1992). If, for example, two time periods are considered, a person might choose to spend all the income which he/she receives in time period 1 within that time period and none of it in the future (period 2) in order to maximize initial consumption. Alternatively, the person could elect to spend and consume less in the first period in order to increase future spending and consumption. These possibilities are illustrated in Figure 2.8.

The person has an income of Y_1 in time period 1 and Y_2 in time period 2 and could choose to spend all of Y_1 on tourism and other goods in period 1 and all of Y_2 on tourism and other goods in period 2. In this case, the opti-

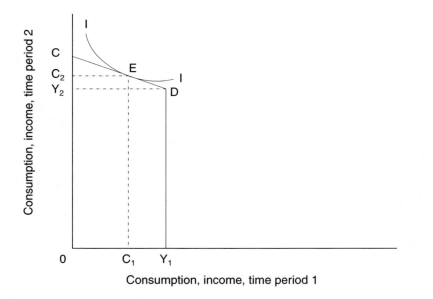

Figure 2.8 Intertemporal choice in tourism consumption

mal consumption point would be at D. On the other hand, the person might decide to consume less in the first period in order to consume more in the second period. In the extreme case of an individual who chose to consume nothing in the first period so as to maximize consumption subsequently, the optimal consumption point would be at C. The range of consumption possibilities which are available from spending less than the full amount of income initially and more subsequently are given by the line CD. The combination of consumption in the two time periods which is chosen depends upon the person's preferences. These may be illustrated by an indifference curve which, in this case, shows the combinations of consumption in each period which provide the same level of satisfaction. In the example in Figure 2.8 the person chose to consume at point E, consuming less tourism and other goods, OC_1, in the first period in order to consume more, OC_2, in the ensuing period.

This case ignores the possibility that people can borrow in order to increase their current consumption or may lend in order to consume more in the future. Introducing borrowing and lending increases the range of intertemporal consumption combinations which are available, as is illustrated by the budget constraint CDC* in Figure 2.9. The line CD is steeper than the equivalent line in Figure 2.8 since a person could receive a higher future income, and hence higher future consumption, via the interest earned from

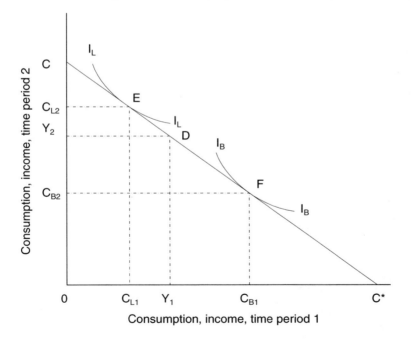

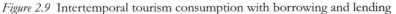

Figure 2.9 Intertemporal tourism consumption with borrowing and lending

27

lending out current income. He/she could also borrow to increase consumption in the first period and the line DC* shows the combinations of (higher) first and (lower) second period income and consumption which could be attained by borrowing. The line CDC* shows all the first and second period consumption possibilities which the person could achieve by lending or borrowing, with OC* depicting the limiting case of maximum first period consumption and zero second period consumption, OC showing the opposite limiting case and CC* the intermediate combinations. The indifference curve $I_B I_B$ illustrates a person who chooses to borrow in order to consume more initially, while $I_L I_L$ depicts an individual who prefers to lend in order to increase his/her income and consumption in the second period.

There are some complications to this analysis because changes in interest rates and inflation alter people's intertemporal consumption possibilities and provide an incentive to consume less now and more later or vice versa. For example, in Figure 2.10, the initial budget line for tourism and other goods is CC*. An increase in interest rates increases the possibilities of future consumption by permitting a rise in future income and decreases the possibilities of present consumption by increasing the cost of borrowing. The new budget line is C'C*' in the figure. The effect of the rise in interest rates may be disaggregated into substitution and income effects. The former is always negative, i.e. there is an inverse relationship, since an increase in interest rates

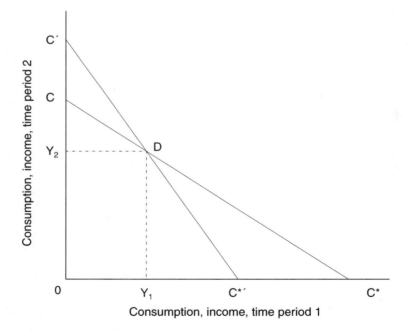

Figure 2.10 Effects of a rise in interest rates on the consumer's budget line

encourages people to substitute (cheaper) future consumption for (more expensive) current consumption. The income effect is positive since the higher income permitted by higher interest rates has a positive, overall effect on current consumption, i.e. the relationship is positive. For a borrower, the net effect of the rise in interest rates on current consumption is negative but for a lender, the net effect could be positive or negative, depending on the relative sizes and therefore net effect of the substitution and income effects. It is also interesting to note that a fall in inflation has similar effects on current consumption to those brought about by a rise in interest rates, since lower inflation results in a rise in real interest rates. The outcome is more complicated if changes in the aggregate rate of inflation are accompanied by a change in the relative prices of tourism and other goods, as additional substitution and income effects have to be taken into account.

THE SOCIAL CONTEXT OF TOURISM DECISION-MAKING

Both the expenditure budget and people's preferences are key variables underlying tourism demand, as shown above. The role of expenditure in determining demand has been examined in the single equation and system of equations models of tourism demand. However, traditional demand theory does not explain how preferences and tastes are formed and change or the process by which decisions are made in the context of the social environment. Research has, instead, concentrated on individual self-interest with regard to the allocation of income to consumption and saving, and the choice of products which consumers purchase.

The emphasis on the individual is reflected in the early tourism literature. For example, propositions by Maslow (1954, 1968), who posited a hierarchy of needs as creating the motivation for the individual to undertake certain activities, have formed the basis of studies in the 1970s and 1980s on the motivation to purchase a holiday and non-economic investigations of motivation have been undertaken by, for example, Gray (1970), Plog (1973), Schmoll (1977), Crompton (1979) and Dann (1981). Motivational studies would benefit from being more strongly focused and although some empirical testing of the hypotheses propounded has been pursued, attention has concentrated mostly on the reasons for and destinations of travel and the marketing implications (D.G. Pearce, 1987).

A common theme running through the theoretical studies of motivation is the need by individuals to escape from their daily work and home regime and to seek new experiences which can be met only by travelling. The direction of research has led to a categorization of forms of tourism and types of tourist. Gray's (1970) 'wanderlust' category suggests a desire to get away, while his 'sunlust' category implies a need which cannot be experienced in the home environment. In his psychographic segmentation of tourists, Plog

(1973) considered the behavioural dimension rather than simply the motivation to travel. He argued that tourists' patterns of behaviour are determined by psychological factors so that the 'allocentric' is more adventurous and self-confident while the 'psychocentric' prefers familiar and reassuring locations and social interactions. Iso-Ahola (1982) largely encompasses earlier theoretical models by Schmoll (1977), Crompton (1979) and Dann (1981) by identifying three principal motivations for travel: escape, seeking psychological benefits in a different physical environment and social interaction. Another branch of research took the form of models incorporating expectations (for example, Rosenberg, 1956; Fishbein, 1963), which investigated the degree to which consumption of particular goods and services resulted in increased levels of individual satisfaction.

These non-economic studies of motivation by social psychologists and, in some cases, geographers make a useful contribution to the development of economic models of tourism demand on two counts. First, they seek to explain the reasons for behaviour which economists observe only from preferences which are revealed in terms of expenditure on goods and services in the market. In this respect, the study of motivation assists in making more accurate explanations and forecasts of the level and pattern of tourism demand. Second, the approaches complement the relatively recent and empirically oriented branches of economics – experimental economics and economic psychology – which have shifted attention towards the social context of decision-making and shed light on the determination of preferences and tastes for tourism consumption. These approaches, while acknowledging the contribution made by other disciplines to explaining tourism demand and behaviour, are set within an economic framework and can help to explain and predict tourist consumption behaviour. The contribution of experimental economics and economic psychology lies in both extending the possible range of variables determining tourism demand and the analytical method, whereby the approach is generally inductive rather than deductive. Moreover, a number of the insights which they have provided have been incorporated into mainstream economic models. Experimental economics emulates the scientific laboratory method by conducting experiments to investigate consumer decision-making and is particularly appropriate for situations in which data relating to key interrelationships are not, otherwise, available. Economic psychology provides a conceptual as well as methodological contribution by recognizing that perceptions, information processing, attitudes, expectations, motivation, preferences and tastes, which are largely unexplained or ill-considered in utility-maximization theory, are important and susceptible to measurement.

Economic psychologists argue that explanation of the process by which decisions are made requires investigation of the social context of decision-making. It is argued that the social environment strongly influences the level of consumption and choice of products at the microeconomic level, as well as the ethos of consumption and saving at the macro level. In his relative income hypothesis, Duesenberry (1949), echoed by Liebenstein (1950), suggested that the level and pattern of consumption of a particular group is determined less by current or future income than by the level and pattern of consumption by other, usually higher income groups. The concept of a demonstration effect implies that consumers look to reference groups and imitate their consumption patterns. Here, economic psychology incorporates ideas from sociology (group theory) and social psychology (attitude postulates), shifting the analytical focus to the microeconomic level.

Support for these views is provided by reference to the historical context of tourism growth. Veblen's (1899) theory of conspicuous consumption and Liebenstein's (1950) snob and bandwagon effect explain how lower-income groups have followed the holiday patterns of the rich. The classification of tourist types, such as the allocentric and psychocentric, and the life cycle of resorts may reflect the influence of social reference groups. The allocentric, innovative tourist seeks new experiences and some less adventurous tourists, generally though not exclusively in lower-income groups, subsequently copy this behaviour. Such changing tastes have led to changes in the patterns of tourism behaviour. For example, demand for some mass-market destinations has decreased, as observed in traditional seaside resorts in the UK, illustrating the phase of decline in a resort cycle. Thus, social factors, including comparisons made by consumers with other consuming units, are seen as important in more psychologically and sociologically based approaches.

Economic psychology and sociology consider economic socialization to be an important determinant of behavioural patterns, preferences and tastes. Economic socialization (Stacey, 1982; Jundin, 1983) concerns the ways in which children are, in stages, introduced to and develop skills with respect to their consumption, from a knowledge of money and possessions to social differentiation and socio-economic understanding. The concept of consumer socialization considers the specific roles of key individuals, for example, parents, in forming children's attitudes to levels and patterns of consumption. The relationship of gender roles and the process of interaction within the family to consumption decisions is also important.

Research has shown that spouses' roles in consumer decision-making differ with respect to the good or service being considered for possible purchase and the stage in the decision-making process at which each partner's views are particularly influential. For most goods and services, the more technically complex the item and the associated information required, the more likely the decision is to be male-dominated (H.L. Davis, 1970; Burns and Ortinau, 1979). However, there has been little investigation of non-durable and durable tourism consumption processes within or between social groups. Regarding holiday decisions, Filiautraut and Ritchie (1980) found that the man's views tended to dominate, but Qualls (1982) found that decisions were made jointly by both partners.

In his review of holiday decision-making, Kirchler (1988) showed that other variables influencing consumption were age, stages in the life cycle, class and income and nationality. In traditional role and older households, consumption decisions were controlled by the husband. In middle-income households, more consumption decisions were made jointly. With respect to social class, decision-making was more male-dominated in the lower and upper classes. The unsurprising conclusion of the few studies undertaken to date is that the person with economic power dominates the initiation, process and outcome of spending but, in changing social and economic circumstances, different people hold this position.

Experimental economics can shed light on the holiday expenditure decision as an interactive process within the consuming unit, as well as between that unit and other social groups, reflecting the social context of tourism demand and behaviour. The methodology can incorporate game theory, in which possible outcomes are determined by the goals and strategies of the participants. For example, one or more members of a household who wish to purchase a beach holiday may collude, while others who seek a more active cultural experience may bargain with them in an attempt to achieve an alternative goal. Kent (1990) used a quasi-experimental approach to examine the process of decision-making, with the added dimension of certain members of the household occupying positions of dominance. Such positions may be achieved and reinforced through the possession of information not available to other members of the group.

Decision-making in situations of risk and uncertainty comes within the purview of both experimental and mainstream economics and could be applied to holiday decisions which involve substantial expenditure, spread over long periods of time, such as the capital outlay on holiday homes. These decisions occur in a context of uncertainty concerning future income, relative prices, inflation and interest rates, and could be investigated using an experimental approach in which various scenarios are constructed and the choices made observed. Some social psychologists argue that even in circumstances in which the consumer could decrease uncertainty by obtaining more information, once an acceptable level of information is attained, the search for and processing of additional information ceases. This may occur well before the consumer has attained the maximum amount of information which he or she can comprehend, and is akin to Simon's (1957) notion of the satisficing consumer. It also accords with the concept of bounded rationality as opposed to fully rational behaviour. The result may be that consumers are not consistent in their decisions, explaining the phenomenon of preference reversal, also indicated by the findings of system of equations models of demand discussed in Chapter 3 and which sometimes occurs in game theory simulations of consumer choice.

In the context of international tourism consumption choices, the employment of the experimental method has questioned the assumptions of

32

models in which expenditure is deferred, for example, the assumption that choices made in one time period are independent of those made in another (Loewenstein, 1987). It, thus, supports the case for analysis of the intertemporal nature of decision-making (discussed on pp. 26–9). Additional investigation of the ways in which individuals and groups obtain, select and use information in their consumption decisions has been undertaken within the field of marketing (for example, Middleton, 1988; Kotler, 1991; Kotler *et al.*, 1993). Increased use of the insights provided by other disciplines can enrich economic theories of tourism consumption decisions. For example, at the micro level it has been hypothesized that tourism demand depends on the economic power of particular individuals and/or previous consumption by social reference groups. It may be possible to estimate quantitative measures of such variables and to include them in the tourism demand function. This would not only correct for omitted variables bias, but also permit the calculation of the magnitudes of the effects of these variables on demand. It can be concluded that the issue of preference formation for tourism consumption, which is particularly relevant to time series analysis of tourism demand, requires considerable further investigation.

CONCLUSIONS

This chapter has considered the conventional microfoundations of tourism demand. One of the reasons why it was useful to do so is that appropriate modelling at the macroeconomic level requires some understanding of the microfoundations and it is at the national, macroeconomic level that most tourism demand estimation has been undertaken to date. Macroeconomic modelling of tourism demand has incorporated, for example, assumptions about separability between consumption and labour supply decisions and the intertemporal separability of consumption. The discussion in this chapter suggests that these assumptions may require modification, indicating that decisions to purchase tourism may be taken in conjunction with labour supply decisions (Figure 2.1) and may also be interrelated over time (Figures 2.8, 2.9, 2.10). It has also indicated that decisions to purchase tourism are related to decisions to consume other goods and services (Figures 2.2, 2.5, 2.6, 2.7) and that choices of particular types of tourism are made in the context of the possibility of purchasing other types (Figures 2.3, 2.4). If tourism demand is related to other types of consumption both at one point of time and over time, such interrelationships should be taken into account in tourism demand modelling, although most past studies have not done so.

Further issues arise in relation to the estimation of tourism demand models. A particularly important one is that of aggregation, which concerns the conditions under which relationships which hold at the individual level apply at more aggregate levels of groups or nations. For example, different individuals and groups have different consumption preferences, so that there are difficulties in identifying preferences at more aggregate levels. This raises problems concerning the choice of model to use for estimating demand at the macroeconomic level. The incorporation of inappropriate preferences in the model can produce inaccurate results. For example, some macroeconomic models involve estimating equations which permit aggregation over groups of commodities but which involve questionable assumptions about preferences, such as equal proportionate responses of demand to proportionate changes in income or relative prices. Such models can provide misleading implications concerning individual or group behaviour. There may be additional problems concerning the interpretation of the estimated equations, for example, when there have been changes in the composition of the population, level of unemployment, liquidity constraints and sporadic variations in the purchase of particular goods so that tourism might be consumed in one period but not the next, as in the case of destinations which experience periods of social unrest. It is clear, therefore, that the theoretical discussion of tourism demand has raised a number of issues which are relevant to empirical studies of tourism demand, which will be examined in Chapter 3.

3

EMPIRICAL STUDIES OF
TOURISM DEMAND

INTRODUCTION

The aim of this chapter is to consider the models which have been used to estimate tourism demand in the context of both the theoretical discussion of the previous chapter and developments in consumer demand theory which will be introduced as the chapter proceeds. Empirical studies can help to explain the level and pattern of tourism demand and its sensitivity to changes in the variables upon which it depends, for example, income in origin areas and relative rates of inflation and exchange rates between different origins and destinations. Such information is useful for public sector policy-making and the private sector. However, accurate estimates can usually be obtained only if the theoretical specification of the underlying model is appropriate. Hence, explicit consideration of the consumer decision-making underpinning empirical models is important in ensuring that the estimates which are provided are neither inaccurate nor misleading in their policy implications.

This chapter will explain and critically evaluate two approaches which have been used to model tourism demand; first, the single equation model and, second, the system of equations model. The single equation model which is considered in the first part of the chapter has been used in studies of tourism demand for numerous countries and time periods and posits that demand is a function of a number of determining variables. The estimated equations permit the calculation of the sensitivity of demand to changes in these variables. The discussion of the approach is centred on identifying its strengths and limitations and considers the associated implications for future research. In contrast to the first approach, the system of equations model requires the simultaneous estimation of a range of tourism demand equations for the countries or types of tourism expenditure considered. The system of equations methodology attempts to explain the sensitivity of the budget shares of tourism demand across a range of origins and destinations (or tourism types) to changes in the underlying determinants. In the second part of the chapter, the theory underlying the system of equations model is explained and the model evaluated. Directions for further investigation are, again, suggested.

The third part of the chapter considers the relevance of the two approaches for estimating tourism demand to methodologies for forecasting demand and evaluates alternative forecasting models. By the end of the chapter, it should be clear that a much tighter link between economic theory and empirical studies of tourism demand is necessary.

THE SINGLE EQUATION APPROACH TO ESTIMATING TOURISM DEMAND

The methodology, its advantages and limitations

Tourism demand can be analysed for groups of countries, individual countries or states, regions or local areas. Demand can also be disaggregated by such categories as the type of visit (for example, holiday and business tourism), and the type of tourist (covering nationality, age, gender and socio-economic group). On another dimension, its analysis can be related to particular types of tourism product, for instance, sports tourism or ecotourism or to specific components of the tourism product, such as accommodation and transportation, discussed in Chapters 4 and 5. Throughout this chapter the concept of tourism demand which is examined refers to the entire bundle of services which tourists purchase: transportation, accommodation, catering, entertainments and related services.

The single equation approach involves, first, theorizing the determinants of demand and, subsequently, using the technique of multiple regression analysis to estimate the relationship between demand and each of the determinants. A demand function might be written as

$$D = f(x_1, x_2, \ldots x_n) \tag{1}$$

where D is tourism demand and $x_1 \ldots x_n$ are independent variables which determine demand. The theoretical issue is, therefore, to identify which independent variables should be included in the equation and the functional form (such as linear or log-linear) which is appropriate for the estimation of the equation. The application of the theory is relatively easy, subject to data availability, by means of computer packages which are designed to estimate the relationship between tourism demand as the dependent variable and the independent variables determining it. Such packages can test whether each of the hypothesized independent variables plays a significant role in determining demand and the extent to which each significant variable explains changes in demand.

The single equation approach has a number of advantages. Apart from being relatively easy to implement, it can provide useful information; once regression analysis has been used to examine the relationship between tourism demand and the determining variables, the extent to which a change in any of them alters tourism demand can be quantified by the calculation of

36

the relevant elasticities. For example, the income elasticity of tourism demand for a destination is a measure of the extent to which demand changes as the result of changes in the income of tourist origin areas. The elasticity value can be calculated for different durations of time, thereby showing the difference between the short-run and long-run responsiveness of demand to changes in the variable under consideration. This may be useful for policy purposes, indicating, for example, the time within which any appropriate countervailing policy adjustment should take effect. Elasticity values can be estimated not only at the national level but also for different groups or nationalities of tourist or types of tourism products, which may differ considerably from the values estimated at more aggregate levels. For example, Gray (1970) pointed out that the price elasticity of demand for holiday tourism is higher than that for business tourism, while wanderlust tourists are likely to have a lower price elasticity of demand than sunlust tourists.

The approach is also subject to various limitations. If the tourism demand equation is mis-specified because of the inclusion of inappropriate variables or the omission of appropriate ones, the results obtained from estimation are likely to be spurious and can lead to inappropriate conclusions and policy recommendations. It may also be the case that, in some circumstances, the tourism demand equation should be estimated in conjunction with the tourism supply equation and/or labour supply equation. Failure to estimate the equations simultaneously may also produce biased results. These considerations further reinforce the case for a strong theoretical basis for the specification and estimation of tourism demand.

Single equation models of tourism demand

Most economic studies have used the single equation methodology to explain tourism demand, usually at the national level, as demonstrated by the reviews by Archer (1976), Johnson and Ashworth (1990) and Sheldon (1990). An example of a tourism demand function, where all variables occur at a given time period, t, is

$$D_{ij} = f(Y_i, P_{ij/k}, E_{ij/k}, T_{ij/k}, DV) \tag{2}$$

where D_{ij} is tourism demand by origin i for destination j, Y_i is income of origin i, $P_{ij/k}$ is prices in i relative to destination j and competitor destinations k, $E_{ij/k}$ is exchange rates between i and destination j and competitor destinations k, $T_{ij/k}$ is the cost of transport between i and destination j and competitor destinations k, DV is a dummy variable to take account of special events such as sporting events or political upheavals.

Studies have used a variety of other specifications of the tourism demand equations resulting in a wide range of estimated elasticity values. For example, an early study of US and Canadian outbound tourism (Gray, 1966) found that the per capita income elasticity values ranged between 4.99 and 7.01, implying that a rise in income of 1 per cent results in an increase in tourism expenditure of between 4.99 per cent and 7.01 per cent. Artus (1972) found that the income elasticities of demand by European tourists for international tourism varied between 1.36 (Switzerland) and 3.84 (Austria). Witt and Martin (1987) found that the per capita income elasticity values for UK travel to European countries ranged between 0.34 (Cyprus, Gibraltar, Malta) and 2.91 (Netherlands) for independent air travel, and between 0.86 (Spain) and 6.35 (Greece) for inclusive tour air travel. The income elasticities for tourism in eighteen European destination countries estimated by Tremblay (1989) ranged between 0.33 (UK) and 11.35 (Portugal). J.S. Little's (1980) study of US demand in ten destination countries found income to be an insignificant determinant of demand. Many more studies have provided additional estimates for a range of origins and destinations, as shown by Archer (1976), Johnson and Ashworth (1990) and Sheldon (1990).

In part, the variations in the elasticity values estimated in the different studies of tourism demand are not surprising, since the values usually refer to different origins and destinations, time periods and measures of demand, such as tourism revenue per capita or visits per capita. However, a number of the estimated values may be inaccurate, owing to inappropriate specifications of the demand equations upon which they were based. In particular, the testing of demand theories has relied heavily on secondary data and econometric approaches in which the random error term contained the variables unexplained by the model. Thus, a central feature of much tourism demand modelling is that some of the variables influencing consumer behaviour have been ignored. Moreover, it is rare for studies to include the full range of test statistics, despite the provision of such statistics in computer regression packages. Thus, the degree of confidence which may be placed in the estimated results is uncertain. In some cases, results which the test statistic indicate as insignificant have been discussed as though they have been of equal importance as those which were shown to be significant, as pointed out by Johnson and Ashworth (1990). In other cases, such econometric problems as heteroscedasticity (where the errors in the regression equation do not have a constant variance) have been ignored.

In the light of the theoretical discussion in Chapter 2 and recent developments in the macroeconomic literature relating to consumption, variables which are likely to be significant determinants of tourism demand will now be considered. Income and intertemporal consumption theory are examined initially. The discussion is then extended to take account of relative prices and exchange rates, which are important in determining the effective relative price of tourism at the international level. Additional variables which may be relevant in the tourism demand equation, notably transport costs, marketing expenditure and dummy variables for special events, will also be considered.

Income and the intertemporal demand for tourism

The analysis in Chapter 2 showed that changes in income can have important effects on tourism consumption, so that it is clearly a key variable within a tourism demand equation. The discussion also examined the theory underlying the timing of tourism consumption. However, a key problem with most studies of tourism demand which have used the single equation model is the absence of an explicit theory of consumer decision-making. Consequently, an explanation of the process by which tourism demand occurs over time and the role of income within it has not been provided. With few exceptions, notably Syriopoulos (1995), the majority of studies have assumed that demand depends on current income but not on past or expected future income. By making this assumption, such research has ignored the debates within the economics literature concerning intertemporal decision-making and the issue of whether consumers are backward-looking or forward-looking. Therefore, in those studies which have assumed demand to depend solely on current income, the relationship between tourism demand and its determinants may have been mis-specified and the calculated elasticities may be inaccurate.

Intertemporal choice theory can help to explain tourism demand at both the macroeconomic and the microeconomic levels. The theory can take account of the fact that demand decisions are often made in the context of imperfect information, unforeseen events, expectations about the future and liquidity constraints which limit current consumption. The theory of intertemporal choice allows consumption to depend on any combination of current, future and/or past income, so that the assumption that it depends solely on current income becomes a special case within a more general model. Intertemporal choice theory argues that people decide how to allocate their consumption between present and future periods so that those with preferences for high current relative to future consumption are said to have a high rate of time preference while the reverse is true for deferred consumption. The value of consumption in any given time period depends on the rate of time preference, the value of current income receipts and possibly on past and expected future receipts. The explanation of current consumption requires the formulation of a theory of expectations with respect to future income receipts because, if people anticipate a change in their income in the future, they are likely to alter their current consumption. In this case, consumers are 'forward-looking' and actual changes in consumption are a function of predictable changes in future income. The implication is that in order to understand changes in total consumption and its tourism component, it is necessary to explain (or model) future changes in income.

A range of theories of expectations of future income have been proposed. Much research on consumption has assumed that expected changes in future income are based on past changes in income. For example, it has been postulated that expectations of changes in income accord with an adaptive process, whereby expectations are adjusted in relation to the past values of consumers' income. Thus, consumption may be a function of the discounted present value of expected future income, predicted on the basis of past values of income, so that consumers are backward-looking. Alternatively, consumers may anticipate that their future incomes will change in a way which differs from past patterns because of what are called innovations in income, for instance, changes in the benefit and tax system. Such innovations may affect expected future income and, hence, its discounted present value, leading to a change in consumption if they persist. Innovations in income vary within and between different countries and time periods and may be incorporated into anticipated future income differently because of varying expectations and contexts.

It is also necessary to take account of the differing amounts of information which are available to consumers in a particular context. The theory of rational expectations, from new classical macroeconomic theory (Sargent and Wallace, 1976), argues that consumers obtain and use all available information when making decisions, so that actual outcomes are consistent with predicted outcomes. According to the theory, unpredictable changes in income lead consumers to revise their expectations about future income, causing changes in the discounted present value of their income and corresponding consumption changes. However, empirical evidence questions this theory of expectations formation (for example, Flavin, 1981; Blinder and Deaton, 1985; Jappelli and Pagano, 1989; J.Y. Campbell and Mankiw, 1991). One possible reason is that consumers are imperfectly aware of innovations in incomes, so that their consumption response is small. New Keynesian macroeconomists posit that consumers' actions may be limited by market imperfections. For example, rigidities in wages may prevent individuals who wish to work from obtaining employment, thereby limiting their income. Some consumers may be unable to increase their current consumption owing to liquidity constraints associated with restrictions on their ability to borrow against their future income. Borrowing constraints may occur because of asymmetric information, whereby the consumer has knowledge of a future increase in income but the lender is unaware of the future increase or is doubtful that it will occur. Uncertainty about future incomes may also reduce consumption by inducing an increase in precautionary savings to cater for the possibility of future decreases in income.

The majority of studies of tourism demand, which have taken account of current but not past or expected future income in the demand function, are consistent with liquidity constraints on borrowing. Such studies are also consistent with consumer behaviour which is neither forward-looking nor backward-looking. Thus, the dependence of tourism expenditure solely on current income can be considered as a particular case within the range of consumer behaviour which can occur. If tourism consumers are backward-looking, the tourism consumption equation should include lagged values of income, weighted according to the degree to which current consumption is determined by past income, with proximate years usually having the highest weights. If consumers are forward-looking, in the absence of liquidity constraints, the tourism consumption function should include terms to take account of the process by which consumption changes in accordance with consumers' expectations of the discounted present value of their future income. This requires information concerning income changes over time and the ways in which people formulate their expectations and incorporate available information into their predictions of future income. Thus, the foregoing discussion indicates the ways in which the tourism demand function can be specified to reflect the relationship between demand and income which is particular to each origin country. The main issue now is to test the theories empirically.

Relative prices, exchange rates and tourism demand

The examination of tourism demand, so far, has concentrated on the ways in which income can affect demand. However, Chapter 2 demonstrated that tourism demand depends not only on its own price but also on the prices of other goods and services, while the choice of different types of tourism also takes account of their relative prices. Moreover, tourism can be either a substitute for or a complement to other goods. Although tourism demand equations should incorporate these relative price variables, they have not been taken into account in empirical studies. First, price indices for tourism are often unavailable so that the retail price index has usually been used although, with the exception of the study by Martin and Witt (1987), virtually no research has been undertaken to investigate its suitability as a proxy. Second, most research has overlooked the possibility that decisions to purchase tourism may be taken in conjunction with decisions to purchase other goods and so the prices of the latter have not been taken into account. Third, many studies have either ignored the fact that consumers choose between a range of tourism products and destinations, or have included the prices of a range of alternatives without providing a well-argued rationale for the range selected.

With regard to international tourism, since the consumer's country of origin is a possible site of demand, the exchange rates between the origin and a range of other destinations are also likely to be relevant. Relative prices and exchange rates between tourists' origin country and their destination have been included in studies of tourism demand, often as separate explanatory variables (for example, Artus, 1972; J.S. Little, 1980; Loeb, 1982; Quayson and Var, 1982; Martin and Witt, 1988; C.K. Lee *et al.*, 1996) but sometimes in the form of effective exchange rates (nominal exchange rates adjusted for differences in relative inflation rates). The latter may be more appropriate in the long run (Syriopoulos, 1995). Occasionally, the prices and exchange rates of other competing destinations have also been incorporated.

As in the case of the income elasticities, studies have yielded a wide range of elasticity values for the relative prices and exchange rate variables included in the estimating equations. For example, the price elasticity values for tourism from the UK ranged between -0.23 (Austria) and -5.60 (Greece), and for tourism from West Germany ranged between -0.06 (Spain) and -1.98 (France) (Martin and Witt, 1988). Artus (1972) estimated price elasticities of European receipts from international travel of between -0.37 (Sweden) and -4.95 (Netherlands). More recently Tremblay (1989) estimated exchange rate elasticity values of between 0.63 (West Germany) and 4.60 (Portugal) for tourism receipts by European countries. US demand for tourism was also associated with a range of exchange rate elasticity values, varying between -0.58 (Mexico) and -3.15 (Canada) as shown by J.S. Little (1980), while Loeb (1982) estimated values of between 0.8 (Italy) and 4.07 (UK). However, relative exchange rates were insignificant determinants of tourism demand in Okanagan (Quayson and Var, 1982) and Singapore (Gunadhi and Boey, 1986). The range of elasticity values estimated may be due to the markedly differing circumstances of the origins and destinations under consideration but, as in the case of the income elasticities, some may result from inappropriate specifications of the equations which were estimated.

In general, there has been little discussion in the literature of whether it is theoretically sound to include relative prices and exchange rates as separate determinants of the demand for tourism at the international level or whether effective exchange rates are appropriate. It can be argued that, over the short run, the rates of inflation and of changes in nominal exchange rates differ, so that tourists take account of relative prices and exchange rates separately in their decision-making. A counter-argument is that tourists are unaware of inflation rates in overseas destinations and take account only of nominal exchange rates. Alternatively, effective exchange rates may be the relevant explanatory variable. One justification for this view is that most tourists pay for their tourism consumption in their own currency and the prices which they are charged take account of both differences in relative prices and exchange rates. Thus, the prevailing methods of pricing and paying for tourism consumption are key considerations as to which variables to include in the estimating equation.

Lagged variables

Studies of tourism demand have usually included the current values of relative prices and/or (effective) exchange rates as independent variables in the estimating equation. However, given that tourism purchases are made in advance of their actual consumption, lagged rather than, or in addition to, the current values may be the appropriate independent variables, depending upon the pattern of tourism consumption in the country in question. Expectations of future changes in prices and exchange rates are less likely to be significant determinants of demand than past rates, given most consumers' lack of information and uncertainty about future movements in them. Consumers' awareness of prices and exchange rates for other destinations is likely to vary between different origins and destinations, and possibly over time, so further investigation of the relationship between tourism demand, relative prices and exchange rates is necessary. This could encompass such aspects as the possibility that consumers are subject to money illusion with respect to foreign currency prices, in that they may be unaware of their domestic currency values in the short run. The possible effects of changes in interest rates in altering the timing of tourism purchases, outlined in Chapter 2, could also be subject to empirical study.

It is likely that the demand for tourism in a particular destination in a given period depends upon demand in the previous period. This is because the demand for tourism in an alternative location may be deterred by consumers' lack of experience and sometimes knowledge of it. Thus, it is commonly assumed that the more information consumers have about a destination, the greater the demand for it. The effect of increased information can be taken into account in the estimating equation by including a lagged dependent variable whereby demand in the current time period is affected by the previous level of demand. This is also consistent with the hypothesis that some consumers develop the habit of making repeat visits to particular destinations (Witt, 1980; Witt and Martin, 1987; Martin and Witt, 1988; Darnell et al., 1992; Syriopoulos, 1995) and is similar to the effect of habit persistence (time non-separability) in aggregate consumption expenditure (Braun et al., 1993). Habit may be explained by the discussion within the tourism literature of the psychocentric tourist, who has a preference for familiarity as opposed to new experiences and destinations; hence, there is a positive relationship (coefficient) between current and past demand. Conversely, a significant negative coefficient may indicate that tourists are allocentric, seeking new experiences in new destinations. Another explanation of a negative coefficient is that previous visits revealed some undesirable feature of the destination. With respect to the lags in the adjustment process, the psychocentric tourist may exemplify adjustment with a long lag or, indeed, no adjustment at all, whereas the allocentric may exhibit very rapid adjustment.

Clearly, the responsiveness of current demand to previous demand may change over time. For example, consumers may initially increase their demand for a destination, owing to greater information about it and/or to the acquisition of a habit of visiting it, resulting in a positive coefficient. However, consumers may subsequently decrease their level of expenditure on the destination owing to increased knowledge of the most cost-effective ways of obtaining products and services within it, for example, cheaper forms of local transport or cheaper restaurants and hotels, for a given level of quality (Godbey, 1988). This would result in a negative coefficient. The possibility of habit persistence and wider information availability have rarely been tested in empirical studies but were found to have a significant positive effect on tourism demand by (former) West German and UK tourists (Witt, 1980; Witt and Martin, 1987) and for south Mediterranean countries (Syriopoulos, 1995).

Transportation

The price of transportation is another variable which some studies have included as an independent variable within the tourism demand equation. The case for and against doing so is complicated. On the one hand, the retail price indices which have usually been included in practice do not take explicit account of the price of transport between the origin and the destination, so that there is a case for the inclusion of a separate transport price variable. Moreover, its cost is such a significant proportion of the total price of a holiday that changes in it may induce a change of mode. On the other hand, the definition of tourism demand which is usually explained is the total bundle of tourism components which are purchased (accommodation, entertainments, other service provision and transportation), so that the own price variable which is included in the estimating equation should take account of the price of all these components. Therefore, the inclusion of a separate price of transport should not, in theory, be necessary. Moreover, even if a case can be made for the inclusion of a transport price variable for a particular set of origins and destinations, the form in which it might appropriately be included is not evident. Nor is it clear which other destinations might be substitutes for or complements to a given destination and, hence, which transport prices might be considered in addition to that for the particular origins and destinations under study. A further issue concerns the specific transport prices which are appropriate candidates for inclusion since, within as well as between most forms of transport, there are different fares, which vary according to such criteria as the pre-booking time, times of travel and length of stay. The differences between scheduled and charter or 'bucket shop' fares for airlines are obvious examples, reliable time series data for the latter being unavailable.

Given these considerations it is, perhaps, not surprising that the transport price variables which have been included in studies of tourism demand have often been insignificant, as in the case of US travel to a range of countries (J.S. Little, 1980; Stronge and Redman, 1982) and tourism in Okanagan (Quayson and Var, 1982). Other studies have found a significant negative relationship, for example Kliman's (1981) study of tourism demand in Canada where the elasticities ranged between −0.94 (Italy) and −3.09

(Portugal) and Tremblay's (1989) study which found values varying between −0.48 (Belgium) and −4.17 (Sweden). Overall, it is apparent that consideration of the price of transport as a possible determinant of tourism demand should be treated with far more caution and be the subject of far more detailed theoretical and empirical investigation than has been the case to date.

Other variables

Other variables which have been hypothesized as determining tourism demand are expenditure on marketing and dummy variables reflecting atypical events, such as sporting attractions or major political changes. Studies of Barbados by C.D. Clarke (1981) and Turkey by Uysal and Crompton (1984) found marketing expenditure elasticity values to be significant but less than unity. The inclusion of dummy variables for special events does not pose such problems and dummy variables for the Olympic Games and EXPO 67 in Canada were found to have relatively small values of 0.35 for the Olympics (Loeb, 1982) and 0.49 for EXPO (J.S. Little, 1980). Political events can have greater effects on demand, as in the case of tourism demand by Indonesians for Singapore when tensions between the two countries were associated with an elasticity of −1.5 (Gunadhi and Boey, 1986).

Implications for research on single equation models

The earlier discussion of single equation models of tourism demand has highlighted a number of issues for future research. These can be classified under the broad headings of theory, aggregation and estimation. The theoretical discussion of Chapter 2 indicated the importance of examining the relationship between total consumption, paid work and unpaid time, on the one hand, and between tourism consumption and the consumption of other goods and services, such as consumer durables, on the other. It also demonstrated the role of income and relative prices in tourism consumption. Empirical research to date has failed to incorporate a thorough examination of any of these issues. It has generally been assumed that tourism consumption is separable from decisions to engage in paid work or consume other goods and services. Moreover, the nature of the intertemporal relationship between tourism consumption and changes in income and effective prices has received little attention. The role of expectations has also been neglected.

A number of studies of demand have included independent variables in the estimating equation on a fairly ad hoc basis. This procedure may have resulted in mis-specification of the equation and biased results. One solution which has been suggested as a way of overcoming bias stemming from the omission of relevant independent variables is the 'general to specific' methodology proposed by Davidson *et al.* (1978), Hendry and Mizon (1978) and Hendry (1983) and applied to single equation models of tourism demand by Syriopoulos (1995). The methodology involves the inclusion in the estimating equation of all possible relevant independent variables and the subsequent exclusion of those that are insignificant determinants of demand. Differences between the short- and long-run responsiveness of tourism demand to its determinants can be taken into account by the inclusion of lagged (previous period) values of a number of the independent variables and the use of an error correction mechanism, so that the short- and long-run elasticity values can be calculated.

The issue of aggregation is particularly important and has not been tackled by empirical studies which have estimated the demand for tourism as an aggregate commodity, for example, UK demand for holidays in Spain, without considering the demand for the components of which it consists or the demand by specific groups of consumers such as families or elderly people. The general failure to consider the nature of the tourism demand equations for different types of tourism products or for different individuals or groups means that there is virtually no evidence about the microfoundations of aggregate tourism demand equations. Therefore, it is not possible to ascertain whether the demand equations which have been estimated were appropriate for the case under consideration and, hence, whether the estimated results were accurate. The conclusion is that the issue of aggregation merits considerable further attention.

Given appropriate specifications for the tourism demand equations, the problem of possible econometric unreliability of results is relatively easily resolved by means of the inclusion and examination of test statistics. However, biased results could also result from the estimation of a single equation for tourism demand in contexts in which the simultaneous estimation of a tourism supply equation and/or a labour supply (paid work) equation is also relevant. If demand and supply are interrelated but this is not taken into account within the estimation process, the econometric problems of identification and simultaneity arise and result in incorrect estimates. An identification problem could occur because, for instance, of supply constraints, such as a shortage of accommodation or aircraft seats. In the context of supply limitations, it is necessary to consider whether demand and supply are interdependent. If so, the demand and supply equations should be estimated simultaneously. Overall, it is clear that there is scope for further investigation of tourism demand, based on rigorous theoretical examination of the appropriate forms of the equations to be estimated, including their dynamic structure which would show the way in which tourism demand changes over time. Such investigations would enable tourism demand equations to be tailored to suit the specific circumstances of the case under consideration.

A final issue which requires consideration is that of expenditure on tourism-related durable assets, for example, a holiday home, timeshare, caravan or boat. Holiday homes or timeshare differ from most durable goods in that they are likely to appreciate or, at least, retain their real value over time and, thus, have a significant effect upon the consumer's wealth. Such purchases, therefore, might affect the future pattern of income and expenditure. For example, ownership of a second home will tend to commit a consuming unit to continuing tourism expenditure in the location of the residence. Moreover, the property may be used as collateral for further borrowing and expenditure, not necessarily related to tourism. The role of tourism property as a particular form of durable has not been investigated although, by implication, this type of asset has been embodied in wealth, thereby influencing consumption (Caballero, 1993). Empirical research incorporating an explicit theoretical framework could investigate, for example, the determinants of changes in the stock of tourism-related durables and the possible existence of cyclicality in expenditure on them. Estimation of the changes over time in the value of the stock of tourism durables, for different areas, would provide useful basic data series.

SYSTEM OF EQUATIONS MODELS OF TOURISM DEMAND

System of equations models of tourism demand have been used to estimate the demand for tourism in a range of destination countries by consumers from a range of origin countries. The models have a strong theoretical foundation, having been formulated using microeconomic theories of demand. The objective has been to develop a model which permits generalization on the basis of the behaviour of an individual who is representative of normal behaviour, so that the equation which is used to estimate tourism demand by consumers in aggregate has an appropriate grounding in individual behaviour. In traditional economic theory, the individual is normally viewed as a 'rational economic person', who is assumed to want more material goods and to behave in an optimizing way to maximize his or her own utility. The individual makes decisions within the framework of a market, within which it is assumed that prices adjust to eliminate excess demand and supply. System of equations models of tourism demand generally assume that individuals make decisions according to the 'axioms of consumer choice'. These state that an increase in price results in lower demand (negativity), the sum of individual expenditures is equal to total expenditure (the adding-up condition), a proportional change in expenditure and all prices has no effect on quantities purchased or the budget allocation (homogeneity) and the consumer's choices are consistent (symmetry). If the axioms are valid reflections of behaviour at the individual level, generalizations to the aggregate level are more likely to be appropriate, permitting the specification of an estimating equation for aggregate demand which, rather than being ad hoc, can be justified in terms of economic behaviour.

47

According to system of equations models of tourism expenditure allocation, decisions are made by a 'stage budgeting process'. The consumer first allocates his/her budget among broad groups of goods and services such as tourism, housing and food, and then to sub-groups, for instance holidays in Europe, the USA and other regions of the world as sub-groups of tourism, followed by allocation to individual items, for example, different countries as holiday destinations. One model commonly used to estimate the allocation of consumer expenditure between a range of goods and services or between a range of countries is the Almost Ideal Demand System (AIDS) model developed by Deaton and Muellbauer (1980a, 1980b). The model incorporates the axioms of consumer choice and a stage budgeting process. The allocation of expenditure between the items under consideration, such as different tourist destinations, may be calculated by means of multiple regression analysis, which provides estimates of the sensitivity of each item's share of total expenditure to a number of independent variables, particularly prices. A typical equation might be

$$w_i = \alpha_i + \sum_{j=1}^{n} \gamma_{ij} \log p_j + \beta_i \log (x/P) \tag{3}$$

where w_i is the share of the budget of residents of origin j allocated to tourism in destination i, p_j is the price level in origin j, x is the budget for tourism expenditure by residents of origin j, P is a price index taking account of prices in the destination areas, Σ denotes the sum of, and α_i, γ_{ij} and β_i are coefficients. Hence, the model takes account of the role of the expenditure budget and of prices in determining tourism demand, in accordance with the theoretical discussion in Chapter 2.

System of equations models of consumer demand have been used to explain the allocation of a tourism expenditure budget between different destinations (White, 1982; O'Hagan and Harrison, 1984; Smeral, 1988; Syriopoulos and Sinclair, 1993) and different types of tourism expenditure (Fujii et al., 1987; Sakai, 1988; Pyo et al., 1991). The objective of these models is, thus, different from that of the models discussed in the previous section, which were concerned with explaining aggregate tourism expenditure but not its distribution by place or forms of expenditure. The AIDS system of equations model was used to examine the demand for tourism from west European countries (the US, UK, former West Germany, France and Sweden) to south Mediterranean destinations (Italy, Greece, Portugal, Spain and Turkey) (Syriopoulos and Sinclair, 1993). It was assumed that consumers first allocated their budget between a range of aggregated categories of goods and services, all types of tourism consumption being one category. Consumers subsequently allocated their tourism expenditure between major regions of the world and, having decided on their preferred region, allocated their expenditure on different countries within the region. The allocation of expenditure within the south Mediterranean region represents the final stage of this decision process and was estimated using an equation similar to equation (3) above.

The results indicated that tourism expenditure elasticities vary considerably, both between destinations for a given origin and between origins from a given destination, as shown in Table 3.1. For example, tourism expenditure elasticities for tourism from the UK varied between 0.88 (Italy) and 2.65 (Turkey), while the estimated expenditure elasticity for German tourism to Turkey was 1.73. The elasticity values relate to the changes in the origin countries' shares of the tourism expenditure budget and indicate that Turkey, for example, would gain a large increase in its share of the tourism expenditure budget as the result of an increase in the size of the budget, while Italy would receive a relatively small increase.

The values of the effective price elasticities (taking account of changes in both prices and exchange rates) also differed considerably between destinations for a given origin and between origins for a given destination, and are included in Table 3.2. The table provides the uncompensated price elasticity values which take account of changes in the real value of expenditure which result from price changes, since these tend to be most useful for policy purposes. The estimated elasticity values were relatively high for tourism from Sweden, the UK and Germany, indicating that tourism from these countries is sensitive to destination price changes. The responsiveness of demand to price rises in destinations appeared particularly high in Portugal and Greece, followed by Spain, Turkey and Italy, so that price competitiveness is an important determinant of tourism demand in these destinations.

Table 3.1 Tourism expenditure elasticities for south Mediterranean countries

	UK	France	W. Germany	Sweden	USA
Greece	1.05	1.26	1.07	2.08	1.43
Italy	0.88	0.85	1.02	0.91	0.83
Portugal	1.58	1.45	1.01	1.32	1.61
Spain	0.90	1.08	0.81	1.06	0.72
Turkey	2.65	2.40	1.73	2.09	1.75

Source: Syriopoulos and Sinclair, 1993

Table 3.2 Effective price elasticities for south Mediterranean countries

	UK	France	W. Germany	Sweden	USA
Greece	−2.61	−0.27	−2.03	−2.44	−0.87
Italy	−1.59	−0.95	−0.80	−1.82	−0.63
Portugal	−2.81	−1.90	−1.35	−3.17	−3.33
Spain	−1.11	−1.17	−1.82	−1.53	−0.44
Turkey	−0.60	−0.51	−1.67	−1.89	−1.66

Source: Syriopoulos and Sinclair, 1993

Elasticity values were also estimated in Fujii *et al.*'s (1985) application of the AIDS model to different types of tourism expenditure in Hawaii, which provided expenditure elasticity values approximating unity. The compensated own-price elasticity values, measuring the responsiveness of the budget share for a particular type of expenditure, such as food, to a change in the price of food assuming constant real expenditure, were generally less than unity. This indicated that the budget shares were insensitive to changes in prices. The uncompensated own-price elasticity values, which allowed for a change in real expenditure because of the price changes, were not significantly different from zero, showing that price changes had little effect on the budget shares.

Advantages, limitations and research implications

The system of equations model has the advantages of incorporating an explicit theory of the consumer decision-making process and of being formulated in a way which is consistent with aggregation from the individual tourism consumer (representative agent) to the macroeconomic level. The approach avoids most of the charges of biases in the results, stemming from an inappropriate theoretical base. Improvements in the estimation of the dynamics of consumer demand are being incorporated in current research (Blundell, 1991) and should permit the methodology to incorporate intertemporal decision-making. The model provides estimates of expenditure, own- and cross-price tourism demand elasticities. These elasticities are estimates of the sensitivity of tourism demand, as shares of total tourism expenditure, to changes in expenditure on tourism, the price of a particular type of tourism or tourist destination and the prices of other tourism types or destinations which may be substitutes for or complements to the tourism type or destination under consideration, as explained in Chapter 2.

Estimated elasticity values have implications for business strategies and policy-making. For example, low expenditure elasticity values for particular tourist destinations may be a cause for concern in that the benefits from increases in expenditure which occur over the long run will go to alternative destinations. This indicates that it is necessary to examine the reasons for low elasticity values and what, if anything, can be done to render the destination more desirable. High price elasticities of demand, significantly greater than unity, may be a source of concern in countries with relatively high rates of inflation and/or depreciating exchange rates, but could be a means of increasing tourism receipts in contexts in which it is possible to improve price competitiveness. The cross-elasticities of demand, which indicate complementarity or substitutability between tourism destinations or types of tourism, provide information which may be useful for tourism marketing campaigns by both businesses and public bodies. Such campaigns could be undertaken in conjunction with other countries or producers in the case of complementarity.

In the system of equations approach, the AIDS model is generally considered to be the most flexible form for representing consumer preferences. However, it assumes that consumption and paid work decisions are made separately, in contrast to the discussion included in Chapter 2 which showed that such decisions may be taken simultaneously. In comparison with the single equation approach, it is less flexible in that all the estimating equations must incorporate the same independent variables and functional form for the equations to be estimated simultaneously. Thus, for example, particular variables, such as political changes or sporting events which may be relevant to the determination of tourism demand in one country but not in others, cannot be taken into account. Moreover, the lag structure of the equations also has to be standardized. Although the model can be used to test the axioms of consumer choice which are supposed to characterize consumer behaviour, the results commonly indicate violation of the homogeneity and symmetry axioms of consumer choice, thereby casting some doubt on the assumption of rationality of the representative consumer upon which the model is based.

The empirical evidence that consumer decision-making does not always accord with the axioms of consumer choice indicates that some of the assumptions concerning consumer decision-making may require modification. For example, consumers may make decisions on the basis of bounded rationality, obtaining and using only limited amounts of information, or may behave in a way which provides them with a satisfactory rather than maximum level of utility, known as satisficing behaviour (Simon, 1957). However, proponents of the theory of utility maximization could posit an all-encompassing definition of utility, whereby some consumers' means of utility maximization involves spending limited time and effort on collecting and processing information, so that they have self-imposed constraints on their knowledge of all the possibilities available to them. Furthermore, it may be argued that the concept of a rational economic person, maximizing his or her own utility, represents a narrow view of consumer behaviour (Sen, 1979). The concept of rationality may be broadened to refer to a person who takes account not only of personal preferences but also of the preferences of others when making decisions. This broader concept of a 'social person' appears to accord with 'green' tourists, who are concerned about the environment, welfare and culture of the communities they visit. It also encompasses tourism consumption decisions in which individual tourism consumers consider the consumption behaviour of other members of a consuming unit such as the family or an external social reference group, when making their own decisions (Kent, 1991). Decisions may be made within a context of market failure in the form, for example, of asymmetric information whereby some tourism consumers have more information (and sometimes power) than others, some are subject to constraints on borrowing and in which there are non-priced externalities and public goods. Some empirical studies have indicated the significance of differences in constraints, such as the ability to borrow, on the actions of different households within the economy (Hayashi, 1985; Jappelli and Pagano, 1988; Zeldes, 1989).

Empirical research has also indicated that economic relationships which hold, over time, at the micro level can differ from those at the macro level. For example, changes in income are often negatively correlated with changes in their own past values at the micro level (MaCurdy, 1982; Abowd and Card, 1989; Pischke, 1991), but positively correlated at the macro level (Deaton, 1992: 90). Moreover, some explanatory variables which are significant in time series consumption functions estimated at the micro level, for example, household size and characteristics, are usually omitted from empirical studies at the macro level. The use of aggregate data alone may give rise to biased estimates of price and income elasticities although micro-based models are not always superior (Blundell *et al.*, 1993). In general, whereas diversity and difference are an issue at the micro level, they are eliminated by aggregation at the macro level.

Aggregate relationships between tourism consumption and income may differ from the relationship between tourism consumption and the income of individual or groups of tourism consumers not only because of income distribution and related effects (Drobny and Hall, 1989) but also owing to differences in the availability of information. For example, an individual may not know, or respond to, information about changes in macroeconomic policy which affects predictions of income and consumption at the aggregate level. Such differences are a further indication of the problems of generalizations based on the concept of an individual tourism consumer. System of equations models have the advantage of an explicit theoretical framework and can provide a large amount of useful information about tourism demand in so far as their assumptions of a representative individual who behaves according to the axioms of consumer choice are met. However, in some contexts their underlying assumptions may require modification, and improvements in the models' ability to take account of the short- and long-run dynamics of tourism consumption are also required. System of equations and single equation models of tourism demand can hide much of what is interesting about decision-making by individuals or groups of tourism consumers. As Sheldon (1990) has pointed out, there is a need for more investigation of tourism demand at the microeconomic level.

FORECASTING TOURISM DEMAND

Single equation and system of equations models of tourism demand can be used as one of three main methods of forecasting tourism demand and are termed econometric forecasting models, the other two being qualitative methods and univariate and multivariate prediction methods. Forecasts of tourism demand are, of course, of interest to members of the tourism industry, as well as to governments and tourism associations. The econometric approach involves estimating tourism demand equations using the relevant explanatory variables, and forecasting demand by including the likely future values of the variables in the equation. The accuracy of forecasts based on econometric models thus depends upon the underlying models which explain tourism demand so that more accurate forecasts can be obtained by improvements in them. The previous discussions of tourism demand modelling are relevant to forecasting in that they indicate the ways in which the models might be improved. For example, intertemporal theory of tourism consumption indicated that tourism may be demanded by forward-looking consumers who discount the value of their expected future income, which involves modelling their expected income over time. If tourism demand depends on current and/or past income, alternative models of future income may be relevant. The econometric approach to forecasting tourism demand also involves modelling the future values of the other variables upon which demand depends, notably relative prices and exchange rates. The occurrence of some events, such as the Olympics or EXPO, is known in advance and can be included in forecasting equations.

Although it is important that developments in demand modelling are incorporated into forecasting models, the choice of model may be subject to considerable disagreement. This is particularly problematic when alternative models provide very different results and implications. Considering, for example, the role of income in determining tourism demand, it has been assumed that, over time, income is 'trend stationary', meaning that income grows according to a predetermined trend, with deviations from the trend having a constant mean and variance (Lucas, 1977). In this case, innovations in income affect the deviations from the trend and tourism demand does not respond highly to the innovations. An alternative theory is that income is 'difference stationary', so that it grows according to a stochastic trend (C.R. Nelson and Plosser, 1982). In this case, innovations affect the growth path of income, rather than deviations from it, and tourism consumption is highly responsive to the changes in measured income. The two theories have different implications for the nature of the business cycle and the effects of government intervention in the economy. It is not clear which model is more appropriate, owing to the common problem within economics of disproving a theory (Lakatos and Musgrave, 1970). One criterion for the choice of model is to use the data for the case under consideration to estimate the alternative models, and to select that which provides results which are not only economically plausible but also econometrically superior. The forecasts obtained from different models can, subsequently, be compared with the actual data, providing a retrospective measure of the models' validity.

In contrast to econometric models of tourism demand, univariate methods involve predicting tourism demand solely on the basis of the past values of demand, without investigating the causes of the past values. Univariate methods are, therefore, inappropriate for those who require explanations of the level of tourism demand and who wish to know the responsiveness of demand to the likely future values of particular variables or to possible alternative values of them. Non-causal, univariate forecasting methods include: calculation and prediction of the moving average of tourism demand; exponential smoothing; trend curve analysis, involving the projection of the trend of best fit; decomposition methods which take account not only of the trend in the data but also of seasonal and irregular effects and the Box-Jenkins univariate method (Autoregressive Integrated Moving Average, ARIMA). The Box-Jenkins multivariate method and Structural model (Unobserved Components Autoregressive Integrated Moving Average, UCARIMA model), applied to tourism in Spanish regions by Clewer *et al.* (1990), are mainly extensions of the non-causal category of models. However, they involve elements of causality in that variables other than the past values of tourism demand are permitted to influence the forecast values of demand.

Qualitative forecasting methods such as the Delphi approach (Seely *et al.*, 1980; Moeller and Shafer, 1987; Green *et al.*, 1990a) which is discussed further in Chapter 8 or scenario writing (BarOn, 1979, 1983; Schwaninger, 1989) constitute a third approach to forecasting. Qualitative methods incorporate expert opinions of likely outcomes and possible alternative scenarios and are used in the absence of reliable time series data. One of the problems associated with qualitative forecasting is that the approach tends to be subjective and the assumptions upon which the forecasts are based are not always made explicit and justified. Hence, the methodology is, in many ways, the antithesis of that which is used in econometric forecasting models, which pay attention to the economic reasoning which is involved in the assumptions which are made. On the other hand, qualitative approaches may be particularly advantageous when long-term forecasts are required, often in the context of a high level of uncertainty. Comprehensive discussions of univariate and qualitative forecasting methods are provided by Archer (1976) and Witt and Martin (1989).

Forecasting models have been applied to tourism in a variety of locations including Barbados (Dharmaratne, 1995), Hawaii (Geurts and Ibrahim, 1975; Geurts, 1982), Florida (Fritz *et al.*, 1984), Puerto Rico (Wandner and Van Erden, 1980), the Netherlands (Van Doorn, 1984), Spain (Clewer *et al.*, 1990; Gonzalez and Moral, 1996) and a range of west European countries, North America, Japan and Australia (Means and Avila, 1986, 1987; Witt *et al.*, 1994; Smeral and Witt, 1996). Studies which have compared the accuracy of different forecasting methods have indicated that, of the quantitative methods, no one technique is always superior to the others (Martin and Witt, 1989; Witt and Martin, 1989). Univariate methods have sometimes provided more accurate forecasts than econometric models and have the advantage of ease of estimation. Econometric methods, on the other hand, provide information about the causes of the future demand, relating to the likely future behaviour of tourism consumers and the economies of tourism origin and destination countries (Makridakis, 1986). The forecasts obtained from econometric models can be adjusted in the light of changes in the economic circumstances of the countries concerned and may be particularly useful for tourism-related policy-making.

CONCLUSIONS

This chapter, along with Chapter 2, has attempted to show how economic analysis can contribute to explaining tourism demand and to evaluating empirical studies which have been undertaken. Economics has the advantage of a theoretical framework which goes beyond the descriptive categorization of tourist types, a shortcoming of some past work. The theory has the additional ability to identify variables which determine demand and to provide the basis for quantification of the short- and long-run sensitivity of demand to changes in these variables. The discussion has shown how some economic theory has been incorporated within models which have attempted to estimate tourism demand, particularly at the national level of aggregation. However, it has also indicated that much demand modelling to date has been ad hoc, with inadequate microfoundations. In addition, the discussion has argued that empirical studies might benefit from theoretical contributions from branches of economics other than the mainstream. The potential of such theoretical analysis and developments has not yet been fully realized as researchers have tended, so far, to restrict investigation to the effects of measurable variables for which data, though often of a secondary nature, are most readily available and which are amenable to the prevailing economic methodology of econometric modelling. In general, the effects of expectations, information availability and other variables which are not easily quantifiable have been neglected and the social determinants of decision-making have frequently been ignored.

Developments in the analysis of tourism demand have, nevertheless, been occurring via contributions arising from improvements in economic theory. The microeconomic theory of consumer demand is one example of an advance in economic theory, providing a framework for explaining tourism demand which can be applied at both the microeconomic and macroeconomic levels, although the analysis could be broadened to incorporate the social context of decision-making. Hence, there is considerable scope for extending research on demand beyond the current emphasis on modelling demand at the national level. The nature and diversity of tourism demand, within its temporally and spatially specific contexts, requires further investigation. For example, the effects on demand of the nature and amounts of information used by different categories of tourism consumers and the constraints to which they are subject merit further research. Insights from a range of disciplines can aid the formulation of the particular questions to be investigated and hypotheses to be tested which might, otherwise, be overlooked. For example, social psychologists have conducted research into the motivations underpinning tourism demand and sociologists have posited that social relationships, both within and outside the household, influence holiday choice.

It is clear that microeconomic studies of tourism demand could provide many interesting findings about tourism demand by different social groups, based on disaggregate models which are specific to the case under consideration. Economic models of the decision-making process and demand for tourism could be formulated for socio-economic classes, gender, race and age groups. The social context of decision-making could be taken into account, at least in part, by the inclusion in the estimating equation of additional explanatory variables which are significant determinants of demand. Some quantitative information about social preferences could be provided, for example, by investigation of the extent to which the demand for tourism in a given destination by a particular class is determined by the past demand for the destination by the same class or by the past demand for an alternative destination by a higher income class (the 'demonstration effect').

The choice of unit which is used to measure tourism demand is important. For example, demand by households does not reveal interactions which occur within the household and which are likely to be related to inequalities in economic power. For instance, the issue of whether tourism consumption decisions are made on an individual basis, imposed by the wage-earner on other members of a household or negotiated between members of a group, has been ignored in most past studies. Thus, consumers' decisions may be made within a context of social pressures and alterations in the social context can result in changes in the pattern of tourism consumption.

Preferences are likely to vary by gender, class and race and information about the effects of gender differences may be obtained by comparison of results obtained from tourism demand equations for men and women. Women's tourism consumption may be a function of their partner's income in the case of women without paid employment, but of their own or of the joint income in the case of women with paid work, and the presence of children is also likely to be relevant. Models of tourism demand could be extended to investigate differences in the ways in which tourists from different socio-economic classes or races and young and old people allocate their expenditure within the destinations of their choice. The investigation of variations in the demand for tourism by different socio-economic groups over the economic cycle could also be undertaken. In those cases for which data are unavailable, interviews can provide useful qualitative information and can indicate theories which merit further investigation.

Demand can, of course, become effective only when backed by income, and the consumer's budget and level and pattern of tourism demand are crucially dependent upon the underlying distribution of income and wealth. Changes in the distribution occurring via, for example, changes in government expenditure and taxation, cause changes in tourism demand. Microeconomic studies covering different socio-economic groups within the population could shed light on the likely effects of distributional changes and comparative country studies could provide some indication of the effects of alternative distributions.

The analysis of the demand for tourism by social groups could be extended by the estimation of disaggregate econometric forecasting models. As in the case of tourism demand analysis, aggregate models of the nation state need not be the main focus of attention but research could encompass forecasts of demand for different components of tourism by different socio-economic categories, based on specific models of their behaviour. The demand for tourism by the different groups is related to the characteristics of the holidays or business trips which they require and the information and opportunities which are available to them. Analysis of the demand for different components of tourism, including transport, accommodation and other service provision, is, therefore, a key focus for future research. It would help to overcome a number of the theoretical and empirical problems which are implicit in tourism demand studies and which have resulted from aggregating a range of diverse services into the composite item 'tourism'. Chapters 4 and 5 will provide the basis for doing so by examining the different components of tourism, the markets in which they are supplied and the pricing and output strategies of the firms within them.

4

THE THEORY OF TOURISM
SUPPLY AND ITS MARKET
STRUCTURE

INTRODUCTION

Tourism supply is a complex phenomenon because of both the nature of the product and the process of delivery. Principally it cannot be stored, cannot be examined prior to purchase, it is necessary to travel to consume it, heavy reliance is placed on both natural and human-made resources and a number of components are required, which may be separately or jointly purchased and which are consumed in sequence. It is a composite product involving transport, accommodation, catering, natural resources, entertainments and other facilities and services, such as shops and banks, travel agents and tour operators. Many businesses also serve other industrial sectors and consumer demands, thus raising the question of the extent to which suppliers can be considered as primarily suppliers of tourism. The many components of the product, supplied by a variety of businesses operating in a number of markets, create problems in analysing tourism supply. It is, therefore, convenient to consider it as a collection of industries and markets and to examine it using not only the neoclassical paradigm but also other schools of thought. This approach allows the analysis not only to cope with the complexities of the tourism product but also to take account of developments in economic concepts, theories and methods, especially the orientation of supply analysis towards industrial economics and the issues with which it has been concerned.

The main objective in this and the following two chapters is to provide a wider and more advanced exposition and application of economic principles to tourism supply than is offered in textbooks on the economics of tourism. Initially, in Chapter 4, the basic tenets of the theory of the firm concerning output, costs, pricing decisions, revenues, profits and losses are explained within the traditional analytical framework of economic models which consider the competitive structures of the markets within which consumers and producers interact. Economic models of different types of market structure provide explanations of the operation of firms under well-defined conditions, each type of market structure being distinctly identifiable. Although, in reality, such conditions may be approximated to rather than attained, the models are, nevertheless, useful in going beyond mere description by providing explanations of firms' behaviour and predicting the short- and long-run outcomes of different market situations. This facilitates the identification of factors likely to be of importance in tourism supply, particularly with respect to the nature and extent of inter-firm and inter-sector competition and the consequent implications for consumer welfare.

First, the extreme case of a highly competitive structure, perfect competition, is examined. Some modifications to the standard assumptions are then introduced using the concept of contestable markets. This is followed by an analysis of monopoly, the limiting case of an uncompetitive market. The discussion proceeds to the less restrictive cases of monopolistic competition and oligopoly which more closely represent the conditions found in most industries. The value of the theoretical constructs is that they impart rigour to the analysis of real life situations. The characteristics of tourism sectors and markets are then outlined and related to the economic theories of competition already examined. Attention is concentrated largely on the types of market structure which apply to three main components of tourism supply, namely accommodation, transport and intermediaries (travel agents and tour operators). A number of key features of tourism supply are identified, reflecting issues which are investigated in industrial economics, almost all of which act as indicators of market structure. Each feature is examined and illustrated by reference to conditions in tourism sectors and markets. The extent to which the economic models can explain and predict the operation of firms in the different tourism sectors is then appraised. The discussion acts as a link to Chapter 5, in which the interrelationships between tourism firms and sectors, especially the strategies adopted in dynamic situations in which uncertainty prevails, are examined in the light of developments in industrial economics. In Chapter 6, further issues relating to market strategies and structural changes occurring in tourism supply are considered in an international context. Analysis of the environment as a key element of tourism supply is undertaken in Chapters 7 and 8.

ECONOMIC MODELS OF MARKET STRUCTURES IN TOURISM SUPPLY

Perfect competition

The model of perfect competition provides a benchmark, illustrating the limiting case of a market involving a very high level of competition. It is assumed that there is a large number of firms and consumers so that neither producers nor consumers can affect the price of the posited undifferentiated product. It is also assumed that there is free entry into and exit from the market, implying no entry or exit barriers. An example which might approach these conditions is the existence of the numerous producers of snacks, meals and drinks which are sold in streets and on the beaches of tourist destination areas in relatively poor countries, although there may be a spatial separation of such sellers which affects the degree of competitiveness.

The cost and revenue conditions for an individual supplier (firm) in a perfectly competitive market in the long run are illustrated in Figure 4.1. As much as the seller wishes can be sold at the price P. Each additional unit of output (such as a snack or drink) is sold for the same price. Therefore the extra revenue which the producer obtains from selling every additional unit of output, the marginal revenue, is equal to the price and is also equal to the average revenue from selling each unit of output. A cost structure is assumed where both the marginal cost of producing each additional unit and average cost per unit first decrease as output rises and then increase owing, initially, to increasing returns to scale and subsequently to decreasing returns. For example, snack and meal producers purchase a certain amount of fuel for cooking and if they were to produce only one meal, the cost of the meal would be relatively high. Since a given amount of fuel can be used to produce more than one meal, the marginal cost of producing each additional meal declines and the average cost of producing the meals also declines. At some point, more fuel, or even another appliance or cook, is necessary to produce more meals, so that the

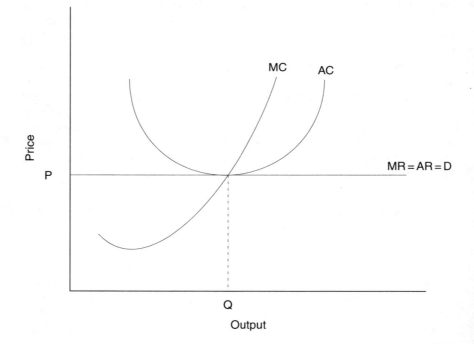

Figure 4.1 Production by a firm in a perfectly competitive market
Note: P = price, Q = quantity of output, MC = marginal cost, AC = average cost, MR = marginal revenue, AR = average revenue, D = demand

marginal and average cost rise. Producers are unwilling to produce for a price less than the prevailing market price, P, as they would make a loss. Hence, the section of the marginal cost curve to the right of Q is the producer's supply curve, since it shows the quantity of output which the producer would like to supply at each price in the short run. Effectively, this means that the quantity supplied will be increased only if prices rise.

The optimal point of production from the supplier's viewpoint occurs where the marginal cost is equal to the marginal revenue, corresponding to an output, Q, in Figure 4.1. This is the point at which profits are maximized, as can easily be demonstrated numerically. Output below Q is associated with a marginal revenue in excess of marginal cost, so that producers wish to increase production because they can increase profits. Conversely, at levels of output above Q, marginal revenue is less than marginal cost so that producers experience reduced profit and would seek to decrease production to the profit maximizing level of output. Hence, the supply curve for the industry is horizontal at the price, P, over the long run. If costs were to increase, for example, fuel became more expensive, producers with higher average costs would go out of business if they are just breaking even at output Q, i.e. earning only normal profits. An overall decrease in costs would result in a short-run outcome of supernormal profits which would attract more producers into the industry until profits returned to their normal level. Thus, within a perfectly competitive market, the prevailing price may result in supernormal profits or losses in the short run. However, there is a tendency towards a break-even price, equal to marginal cost where average cost is at a minimum, and consumers appear to benefit. This raises the issue of whether tourism markets which involve low levels of competition whereby supernormal profits are earned should be rendered more competitive and whether this would increase consumers' welfare. There is no unequivocal answer because, in the context of economies of scale where an increase in the firm's output is accompanied by a fall in the average cost of each unit of production, it is possible for imperfectly competitive markets to be more efficient than those under perfect competition. This issue is taken up below.

Contestable markets

It has been argued that although most real world markets are not perfectly competitive, a significant number give rise to similar economic outcomes. Baumol (1982) introduced the concept of contestability to take account of this outcome. A contestable market is characterized by insignificant entry and exit costs, so that there are negligible entry and exit barriers. Sunk costs, which a firm incurs in order to produce and which would not be recuperable if the firm left the industry, are not significant. Because of reasonably efficient information flows, the same supply conditions and technology are available to all producers. It is posited that producers are unable to change prices instantaneously but consumers react to them immediately.

The key insight of contestability is that new and existing firms find it possible to challenge the position of rivals through pricing strategies. Thus, firms in contestable markets operate similarly to those in perfectly competitive markets in that they charge approximately the same price for a given product. Although economies of scale and scope may arise, incumbent firms are unable to charge a price exceeding average cost because this would attract competitors into the market. Rivals would not be averse to entry because of the low sunk costs and low entry/exit barriers. Hence, contestable markets may be beneficial to consumers. For example, independent tour operators who are not vertically integrated with an airline, accommodation chain or other facilities are governed by many of the conditions prevailing in this type of market, especially ease of entry and exit and minimal economies of scale. N. Evans and Stabler (1995), in considering the UK air inclusive tour market, discuss such behaviour using the categories of 'second tier' and 'third tier' tour operations. Fitch's (1987) and Sheldon's (1986) earlier work also pointed to tiers of tour operations. The same kinds of conditions apply in travel agencies where, notwithstanding the presence of multiples in the sector, at the outlet level operational costs are not markedly below those of independent firms (Bennett, 1993).

Monopoly

Monopoly represents the opposite extreme of perfect competition. Unlike producers in a perfectly competitive market, the monopolist has considerable control over the product price and level of output. Given normal demand conditions, in order to sell additional output, the product price must be lowered so that average revenue falls and consequently the marginal revenue per additional unit of output sold decreases as the amount sold rises. The relationship between the two forms of revenue is such that marginal revenue decreases by twice as much as the average revenue and hence the marginal revenue curve lies below the average revenue curve, as illustrated in Figure 4.2. Following economic suppositions regarding cost structures, returns increase and then decrease; therefore the marginal and average cost curves initially decline and subsequently rise.

Profits are maximized at a level of output where the marginal cost of production equals the marginal revenue and marginal cost is rising (or in cases where increasing returns prevail, marginal cost is falling less rapidly than marginal revenue). As shown in Figure 4.2, the average cost of production is lower than the price charged giving rise to supernormal profits, above the minimum required to keep the firm in the industry. Thus, consumers are paying a price in excess of that which might emerge in a more competitive market. This raises the issue of whether the monopoly should be allowed to trade unhindered or be regulated or reorganized as a competitive industry.

Various components of tourism supply in different countries are deliberately organized as monopolies. For example, domestic air flights may be

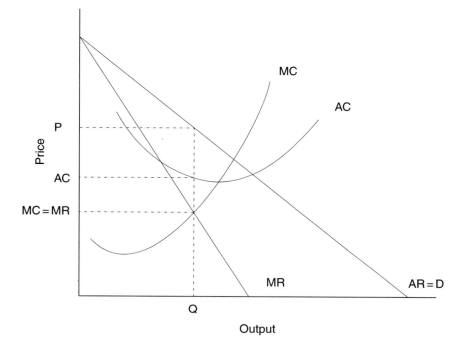

Figure 4.2 Production by a monopoly
Note: P = price, Q = quantity of output, MC = marginal cost, AC = average cost,
MR = marginal revenue, AR = average revenue, D = demand

monopolized by the state airline and railway networks sometimes operate as
a single industry. This may appear paradoxical as it would appear to counter
the interests of consumers. However, two interesting outcomes may occur
under monopoly as opposed to competition, which illustrate the debate concerning the relative advantages and disadvantages of each type of market
structure. The first case involves a comparison of the equilibrium price and
output combinations under competition and monopoly respectively, in which
it is assumed that a competitive industry is monopolized without any change
in production conditions. The outcome is illustrated in Figure 4.3, which
assumes that each competitive firm has the same costs and therefore, for simplicity, the average cost curve can be excluded. It is clear that the price, P_1, is
lower and the quantity produced, Q_1, higher under competition than under
monopoly. If the industry were monopolized, profit maximization would
occur where the short-run marginal cost, SMC, is equal to the marginal revenue, MR, resulting in a higher price, P_2, and lower output, Q_2. In the long
run, the monopolist could close some production units, decreasing output to
Q_3 and increasing the price to P_3. Thus, consumers would be worse off under
monopoly than under competitive conditions. This could occur in the

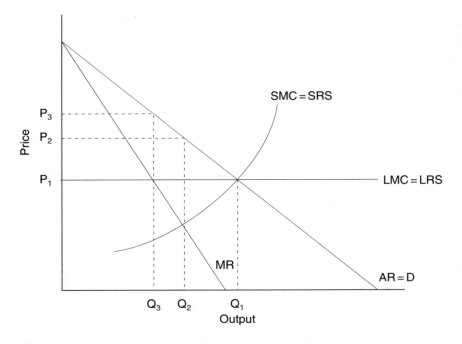

Figure 4.3 Short-run and long-run production by a perfectly competitive industry and a monopoly
Note: P = price, Q = quantity of output, SMC = short-run marginal cost, LMC = long-run marginal cost, SRS = short-run supply, LRS = long-run supply, MR = marginal revenue, AR = average revenue, D = demand

tourism accommodation or intermediaries sectors, but more particularly in the transport industry where small independent airlines, bus, coach or ferry operators are amalgamated as, to an extent, has occurred with deregulation; after an initial influx of new entrants, larger enterprises have taken over many of the smaller firms and exercised monopoly powers.

In the second case, production conditions differ according to whether the industry is operated as a monopoly or under competition as large economies of scale can apply when production is undertaken by a single firm. This applies in the case of a so-called natural monopoly in which both the marginal and average costs are lower over the range of output which could be purchased in comparison with competition. In these circumstances, consumers could benefit in terms of a lower price and higher quantity of production even if the monopolist made supernormal profits. This is illustrated in Figure 4.4, which shows the price and quantity combinations under an unregulated monopoly, P_1 and Q_1. If the monopoly were regulated so that the price charged was equal to the marginal cost of production at P_2, consumers' welfare would increase owing to the lower price and higher output,

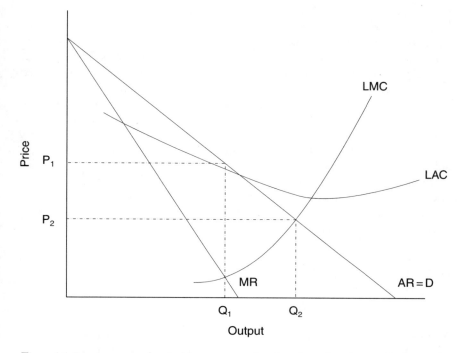

Figure 4.4 Long-run production by an unregulated and regulated natural monopoly
Note: P = price, Q = quantity of output, LMC = long-run marginal cost, LAC = long-run average cost, MR = marginal revenue, AR = average revenue, D = demand

Q_2. However, a government would need to subsidize production since the price would be lower than the average cost of production, so that the supplier makes a loss. For example, most rail systems operate at a loss in terms of not covering average total costs because they are viewed as providing a public service and may be required to charge prices at which they endeavour to cover operating (marginal) costs and contribute to meeting fixed costs. Even if the monopoly were not regulated consumers might benefit, in the long run, from product and process innovations resulting from the reinvestment of some of the profits gained by the monopolist, including the necessary research and development. Under competitive conditions, the availability of funding for research and development is sometimes problematic. Therefore the issue of whether monopolies in tourism supply, as well as in other sectors of the economy, should be allowed to exist or be regulated is complex and varies according to the differing circumstances of particular industries. The absence of a general conclusion concerning the relative merits of the different market structures strengthens the case for specific empirical investigations of the tourism sectors in question.

The difficulty of predicting a determined outcome can be aptly illustrated by reconsidering the examples of state-sanctioned monopoly rail and airline markets. Many governments have partially or totally privatized a number of services, which were hitherto perceived as natural monopolies. The rationale has often been that, notwithstanding economies of scale, they could be operated more efficiently as private sector industries. Privatization has often been accompanied by deregulation to encourage new entrants and increase competition. However, paradoxically, the eventual effect has often been to bring about greater concentration. One important reason is that smaller new entrants are unable fully to exploit economies of scale and therefore fail. It is still too early to establish empirically what the structure of the airline, bus, ferry or rail sectors will be. Effective regulation may be needed to prevent abuses of monopoly power in the form of high prices and the extraction of supernormal profits or restraint of competition. In practice, governments usually define monopoly as occurring when a single producer accounts for a specific, relatively high percentage of production of a particular product and monitor the operations of the firm in question in an attempt to ensure that consumers are not disadvantaged.

Monopolistic competition

Monopolistic competition is a type of market structure, often associated with retailing, which is intermediate between perfect competition and monopoly. It is similar to that of perfect competition and contestable markets in that there is ease of entry and exit in the long run. However, it differs in that suppliers have some control over the price for which they sell their product and, thus, also over their price/output combination and associated market share. However, the pricing and output decisions of an individual supplier do not have a significant impact on those of another because it is normally assumed that there are many suppliers and no substantial degree of concentration. There are also usually only limited economies of scale, unlike the cases of monopoly and oligopoly. In tourism monopolistic competition applies, in many respects, to the hotel accommodation sector which is characterized by many suppliers who provide products which are close but not exact substitutes, so that there is some degree of product differentiation reinforced, as in retailing, by the spatial separation and location of businesses.

In the short run, suppliers within a monopolistically competitive market can charge a price which provides them with supernormal profit. Firms produce where their short-run marginal revenue, SMR, is equal to their marginal cost and charge a price which exceeds their average cost, as illustrated by the price and output combination, P_1 and Q_1, in Figure 4.5. However, in the long run, supernormal profits, combined with the virtual absence of entry and exit barriers for the industry, attract new competitors so that existing firms experience a decline in the demand for their products. This is illustrated by a shift to the left of the average revenue (demand) curve, SAR, and the short-run marginal revenue curve, SMR, until they attain their long-run positions, given by LAR and LMR with an equilibrium price P_2 and output Q_2.

Over the long run, demand decreases until the break-even point where average revenue, shown as the price P_2, is equal to the average cost of production so that there is no further entry into (or exit from) the industry. Although output has contracted and supernormal profits have disappeared, the price charged is still greater than the marginal cost of production. Hence this form of competition appears less efficient than perfect competition although the wide variety of products provides consumers with greater variety of choice. There are a number of examples of tourism businesses in this position such as the smaller firms in the contestable segments of the accommodation and transport sectors, where the extent of their market limits the possibilities of operating at a level which reduces their costs (see Figure 4.5). This might explain why smaller airlines, bus and ferry operators eventually

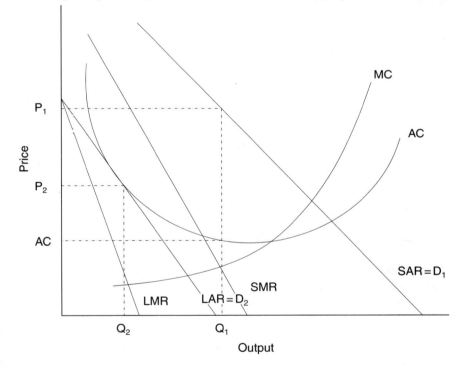

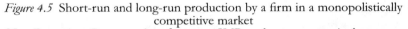

Figure 4.5 Short-run and long-run production by a firm in a monopolistically competitive market
Note: P = price, Q = quantity of output, SMR = short-run marginal revenue, LMR = long-run marginal revenue, SAR = short-run average revenue, LAR = long-run average revenue, MC = marginal cost, AC = average cost, D_1 = short-run demand, D_2 = long-run demand

get taken over as they cannot compete at prevailing prices, especially during a price war.

Oligopoly

An oligopolistic market structure occurs when a small number of producers dominate the industry, the international airline industry being a case in point. Each firm has some control over its price and output decisions and there are some barriers to entry and exit. The key characteristic of oligopoly is the interdependence between producers so that each firm's price and output decisions depend, in part, on those of its competitors. One well-known example of such interdependence, which has become a standard case in economics, is that of the oligopolist's perceived kinked demand curve, which shows the likely outcome for a firm, should it contemplate a change in its price in the absence of changes in cost or demand conditions which affect the whole of the industry. It knows that if it decreases its price its competitors will follow suit, demand for its own product becoming more inelastic, so that it will not increase its market share by doing so. Conversely, if it increases its price they will maintain their own prices so its demand is more elastic and it will lose market share. Thus, the prevailing market price is the profit-maximizing price for the firm. This is illustrated in Figure 4.6.

The equilibrium price and quantity of the oligopolist are P and Q. If the firm increases its price above P, it experiences a large decrease in demand as its competitors fail to decrease their prices in line, so that the average revenue (demand) curve for the oligopolist is relatively elastic (flatter) between A and B. If, in contrast, it decreases its price below P, it increases its sales only marginally because its competitors also tend to decrease their prices, so that its demand curve is relatively inelastic (steeper) between B and C. The firm's marginal revenue curve is, similarly, relatively elastic up to the quantity Q and relatively inelastic for quantities in excess of Q. Thus, the prevailing price, P, and output, Q, tend to be stable in the absence of collusion between oligopolists.

Ideally, individual oligopolists prefer prices to be set at a level which maximizes the joint profits of all producers in the industry. In effect, if all the firms in the industry could act together, the result would be similar to a monopolist, leading to an increase in price and a decrease in the quantity sold, in contrast to the outcome that would occur under competitive conditions, as previously demonstrated in Figure 4.3. The possibility of obtaining supernormal profits by this means provides a rationale for inter-firm collusion. For example, inter-airline pricing and route-sharing agreements have been one example of a strategy which has been used to increase joint profits. On the other hand, there is an incentive for an individual firm to cheat as, if it could do so successfully, it would increase its profits and market share relative to those of its competitors. The stability of a number of pricing and output-sharing arrangements may be

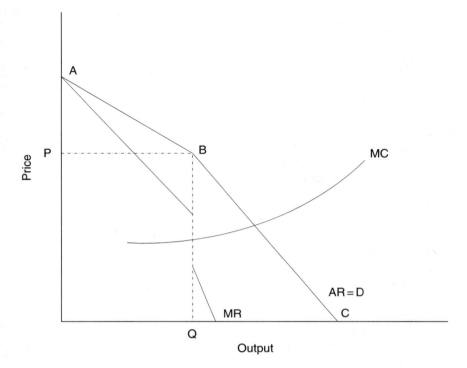

Figure 4.6 Production by an oligopolist
Note: P = price, Q = quantity of output, MC = marginal cost, MR = marginal
revenue, AR = average revenue, D = demand

accounted for by the fact that should all firms cheat, all would be likely to be
worse off. Oligopolists can alter their output as well as the prices charged and
take account of their competitors' possible reactions when they make their
price and output decisions, as was shown in the case of the Bermuda resort
hotels market (Mudambi, 1994; Baum and Mudambi, 1995). The possible
strategies and responses which can occur have been examined by means of
game theory. This is one of the most rapidly developing areas of analysis
within industrial economics and is examined in Chapter 5.

As shown by a number of illustrative examples, the competitive structure
of some aspects of tourism supply can be explained within the mainstream
economic analysis of the theory of the firm. However, to consider more fully
not only the relevance of these basic constructs but also their limitations, it
is necessary to delineate a morphology of tourism sectors and markets and
describe their principal characteristics. This is undertaken in the next section,
which also acts as a link to the following chapters in which current issues are
examined.

MARKET STRUCTURES IN TOURISM SUPPLY

While some reference to tourism market structures has already been made, given the many sectors within tourism supply, it is instructive to consider the extent to which those with similar characteristics can be grouped together within one of the economic forms of competition explained above. In order to do this, it is necessary to identify and describe the nature of supply. The approach adopted in this section is to outline the sectors which have been recognized in the tourism literature. Those which appear representative of the market structures in tourism are described concisely and the key factors indicating their competitive status are identified, such as the degree of concentration, entry and exit conditions, pricing strategies, profit levels, product differentiation, cost structures and capacity and interaction between firms.

The problem with any classification of the supply components of tourism is how broad or narrow they should be. Such categories as transport and accommodation are very broad and benefit from disaggregation into submarkets with different structures and modes of operation. The classification of tourism markets adopted here largely follows the convention adopted in the tourism literature (for example, Cooper *et al.*, 1993; Holloway, 1994), with some modifications to the nomenclature employed (Table 4.1).

Discussion of tourism supply has been conducted largely outside economics, with the result that coverage of economic issues is patchy and there is no coherent overview. Many texts directed at practitioners, particularly in the hospitality sector, touch on supply and associated issues in discussing financial structures, management, marketing, quality and training. At a very specific and applied level, studies of the planning, development, operation and performance of such enterprises as hotels, guesthouses, holiday villages, ski and timeshare resorts and theme parks have been undertaken. At a more general level, attempts have been made to model tourism locations, for exam-

Table 4.1 Major tourism markets

Accommodation	Serviced
	Self-catering
Transport	Air
	Rail
	Road
	• coach
	• car hire
	Sea
Intermediaries	Travel agents
	Tour operators
Attractions	Natural
	Human-made
Other services	Private
	Public

ple by geographers who aim to identify key factors determining growth or decline. With the exception of transport sectors, economists' interest in the industry has been peripheral, largely arising from research into industrial structure or organization, for instance multinational enterprises in the hotel and airline sectors (Dunning and McQueen, 1982a). Economists' tendency to neglect the service sector in general and tourism, in particular, creates difficulties because there are large gaps in the empirical evaluation of parts of the industry. While some reference is made to the other sectors, in order to keep the examination within bounds attention will be concentrated on accommodation, transport and intermediaries as illustrative cases.

The accommodation sector

Cursory inspection of the accommodation sector might suggest that a few large chains dominate the market, giving the impression of an oligopolistic structure. However, the service hospitality sector within holiday tourism is mostly fragmented in many small units where location and the spatial distribution of accommodation are important factors determining the degree of competition. Furthermore, the wide range and quality of accommodation, its multi-product nature (for example, camping, caravanning, holiday centres, timeshare, as well as serviced accommodation) and seasonal variations in demand, introduce an additional dimension into the operation of the market. These aspects of the accommodation sector are considered at some length in tourism texts, of which those by Holloway (1994) and McIntosh and Goeldner (1990) are good examples. Accordingly, different forms of structure – perfect competition, monopolistic competition, oligopoly and even monopoly – might, as argued below, reflect the conditions of different elements of the sector, ranging from the serviced to the unserviced self-catering segments.

Notwithstanding its complicated structure, some fundamental economic factors characterize the accommodation sector. It is subject to fixed capacity with all its attendant problems in the face of periodicity, perishability and seasonality. Allied to this, particularly in larger units offering a wide range of services, high fixed costs drive operators to attain high occupancy rates through such devices as product differentiation and market segmentation. These characteristics tend to involve elements of both natural monopoly and oligopoly. For example, some hotels concentrate on the luxury segment while others serve a budget clientele. Many seek flexibility by targeting the business market during the working week and the leisure sector at weekends. In holiday resorts, the needs of different groups can be met over the year, for instance catering for skiers in winter and walkers in the summer, as occurs in Austria, France and Switzerland. Despite these attempts to exploit market potential, in the early 1990s hoteliers who bought at the peak of the property boom in the UK were put under extreme pressure as they had made an overpriced purchase which could not be serviced by revenue. The property market collapse exacerbated the problem as the fall in property values reduced the collateral security.

There is evidence (Horwath Consulting, 1994) that some forms of accommodation can exploit economies of scale not only within individual establishments but also by the management of a large number of hotels. This partially explains the existence of chains which control many hotels, such as Choice, Accor, Bass and Forte/Granada, the four leading international groups. To an extent it also accounts for concentration in the sector, the nature of ownership and location being other explanatory variables. For example, in the USA with a much higher proportion of corporate ownership, the concentration ratio is over 30 per cent while in eastern Mediterranean holiday resort destinations, comprised of family-owned businesses, it is less than 2 per cent. Some large accommodation groups also attempt to enlarge their market share and control not only by takeovers and mergers but also by franchising, leasing, management contracts and collaborative agreements. Economic integration between firms may result in reduced fixed costs and entry barriers and may make it possible to increase occupancy rates by tapping new segments of the market. Independent and smaller operators sometimes try to combat the power of the large firms by forming co-operative consortia to reduce overheads, for instance by setting-up referral or reservations systems.

Large corporate-run hotels, mainly serving business travellers, tend to cluster in or around large urban areas, airports and on land transport routes. Holiday hotels are more likely to be independent and more widely dispersed, although clustering still occurs, such as in resorts or locations which are principal tourist attractions. In this sense the accommodation sector is akin to monopolistic competition in the retail market, in which accessibility and complementarity, central tenets of urban economics, are characteristics (Balchin et al., 1988). Urban economic theory shows that key locations confer benefits which give a commercial advantage and therefore a higher turnover and profits, explaining why large firms can outbid smaller ones in high cost areas. In effect, a market monopoly can arise because of locational monopoly of specific sites. It also helps to account for the pattern of accommodation in holiday areas like seaside resorts where smaller, lower quality hotels are pushed into secondary locations away from the seafront. This is explained in terms of spatial rent/property price gradients whereby, for individual businesses, location is determined by the property/land costs each faces which are dictated, in turn, by the demand for specific sites by competing activities. Around airports, in urban centres and along seafronts, land costs are high whereas further away they are lower as demand for sites is less. Thus, businesses with lower revenue and profits are forced into areas with lower rents/prices. However, although the intensity of competition is lower in particular destinations outside locations subject to such clustering, it is still significant in terms of the total market, given the number of holiday choices open to consumers and the size and quality range of accommodation available. For example, in the summer sun package market there is a large number and wide distribution of destinations and types of accommodation so that, in the face of static demand, the market structure is highly competitive.

The self-catering sector is an interesting area of the accommodation market which has shown both high growth rates and great diversity. It has complemented the serviced sector by providing additional rooms whose clients are allowed use of the facilities provided, and it has also been incorporated into purpose-built or redeveloped tourism centres offering many attractions and facilities. Holiday centres and villages and timeshare come into this category. There are a number of large players in the field, for example in the UK, Butlins, Haven, Holimarine, Pontins and Warners, formerly low-cost concerns which have upgraded their product, while the Center Parcs organization is an innovative market leader. Disney Vacation, Hapimag, Vacation International, Villa Owners Club and timeshare are similar in terms of the holiday centre concept but differ in that they involve capital investment by consumers. However, none of these control the market which displays two principal economic features. The first is the wide range of standards of quality and therefore segments of the market targeted. The second is its relative infancy, making it difficult to predict its eventual characteristics and competitive structure.

The picture which emerges of the holiday accommodation is of a fragmented sector dominated numerically by small and medium-sized businesses, even if not in terms of the value of total market transactions. This is largely the result of its relatively low entry barriers and generally labour-intensive nature, where the emphasis is on customer service within a small market area, owing to the spatial separation of both demand and supply. Thus, the market exhibits characteristics which reflect more than one form of economic classification, ranging from highly competitive conditions in the small unit sector in clustered locations, to a virtual monopoly of specific locations by large hotels. However, the appropriateness of neoclassical analysis of tourism accommodation is put to the test by its segmentation and by its temporal and spatial dimensions. Consequently, two branches of economic analysis appear to be pertinent to understanding and explaining the accommodation market. Industrial economics' concern with market structure, entry conditions and concentration, product differentiation, segmentation and spatial competition is certainly relevant. Urban and regional economics, with its focus on the determinants of location and spatial distribution of activity, can also make a contribution. These approaches are not examined in detail in this book but are considered in texts such as A.W. Evans (1985) where the theory of office and retail establishment location is discussed, paralleling the principles underpinning the location of the various forms of accommodation.

Overall, taking account of the significance of spatial factors in moderating market conditions, on a continuum of perfect competition to monopoly, the market structure of tourism accommodation tends to accord with the contestable–monopolistic competition positions. In business and resort centres, large hotels experience oligopolistic conditions but outside these areas the structure is closer to monopolistic competition. The picture is clouded somewhat by the intra-sectoral choice of type and quality of accommodation with some segments, for example self-catering, being more contestable than others. The lack of empirical evidence precludes making further inferences.

The intermediary sector

Tour operators and travel agents respectively assemble and retail holidays largely for the mass market. The role of tour operators is to supply holiday packages and to facilitate the link between the suppliers of travel, accommodation, facilities and services, both in origins and destinations, and the tourist. They procure the components of the product, usually by negotiating discounted prices, and retailing it through travel agents or directly to the customer. Tour operations are conducted in a number of ways: by an independent firm specializing solely in holiday assembly and marketing; as a subsidiary of a conglomerate business with diverse interests; as a division of an airline; or linked with a travel agency.

Tour operators

Apart from their principal–agent relationship, an issue of some significance in industrial economics, the structure of the intermediary market, for example, in Europe and the USA, raises other interesting questions. There are some large players in the market in terms of both ownership and control and market share. In the USA, there were approximately 1,500 in total in the early 1990s, of which forty tour operators (3 per cent) controlled almost a third of the market. In the UK, the top ten tour operators accounted for over 70 per cent of the 17 million seats of ATOL (Air Travel Organizers' Licence) licences granted in 1993/94. Four companies – Thomson, Airtours, Owners Abroad (First Choice from 1994) and Cosmos – took almost 60 per cent of the air holidays licences (Civil Aviation Authority (CAA), 1994; N. Evans and Stabler, 1995). In addition to this marked concentration, there is also extensive vertical integration with respect to travel, accommodation, travel agency and holiday services.

Other features of tour operation of note are, despite the degree of concentration, the rate of increase in and the total number of firms. In the USA nearly 600 companies were operating in the late 1970s but this number increased to over 1,000 by 1985 (Sheldon, 1986). Likewise, in the UK there were around 500 firms in 1985 which, by 1993/94, had nearly doubled when 1,000 companies were licensed (N. Evans and Stabler, 1995). Of more significance, however, is the birth and death of tour operators, mostly of smaller firms. It has been estimated that only a third of those in business in the late 1970s were still extant in the middle to late 1980s in both the USA and Europe (Sheldon, 1994). In the UK, twenty-four firms ceased trading in the year 1992/93 and another twenty in 1993/94, notwithstanding which the total number of firms increased.

The performance of tour operators is very sensitive to market conditions, particularly variations in demand arising from such factors as changes in exchange rates, economic recession in origin countries and inflation and perceived political instability in destinations. In the face of tourism development and the continuing extension of possible destinations, capacity has sometimes outrun demand. Furthermore, as in the hospitality market, seasonality is significant. Large firms rely on a high volume of sales with low margins, a significant proportion of profit being generated by investing receipts from advanced bookings. A high rate of sales makes it possible to achieve substantial economies of scale and scope through operating efficiencies, a wide knowledge of an extensive market and market power to gain large discounts from carriers and hoteliers, where past performance is the key. Provided fixed investment is low, as in the case of smaller and more specialized tour operators, the return on fixed investment can be relatively high. Nevertheless, many firms suffer losses over a run of years and when net profits are made they are modest, the return often being less than 4 per cent in the UK, although there is considerable variability.

In the inclusive package market there is intense competition to secure sales volume in order to generate cash flows. Consequently discounting can be widespread at the launch of the subsequent season's packages to encourage early booking. Discounting can also occur at the end of the season to fill excess capacity. The expectation of discounting by potential tourists, who delay making a firm booking earlier in the season, exacerbates tour operators' problems because not only do they have to dispose of holidays at little above cost but also their cash flow is adversely affected. In the early 1990s, late bookings in the UK accounted for almost 40 per cent of the summer sun package market. However, in August 1996 the leading tour operators withdrew holidays and raised prices in an attempt to deter such action by consumers which had led to considerable decreases in profits in 1995. Discounting strategies are both a manifestation of the drive to maintain or increase market share and a reflection of the long lead times, often up to three years, required to launch new holiday types, increasing the likelihood of overestimating demand leading to over-supply. Even large companies indulge in price wars, the leaders in the market often initiating them to safeguard their market position.

A number of factors such as ease of entry and exit, the number of tour operators, fierce price competition, low margins and often significant losses all point to contestable if not highly competitive market conditions in the UK and many other countries. However, the degree of concentration of market share in the package holiday segment suggests an oligopolistic structure. As with the accommodation sector, the reality of the market is more complex than can be encompassed in any single theoretical model of market structure. It appears necessary, therefore, to consider different segments as being characterized by different competitive conditions. This, together with the apparent inherent instability, suggests an immature market which tends to show up the limitations of the neoclassical paradigm.

Travel agents

Travel agents function as brokers in arranging all aspects of business travel and holidays but act as agents in that they represent principals, whether they be tour operators or the ultimate suppliers such as carriers, hoteliers, car hirers and insurance companies. The travel agency component of tourism supply is concentrated in that a limited number of firms with multiple outlets, some integrated with tour operators or carriers, dominate the market, particularly in the USA. There is a need to generate a high turnover in the face of low margins; commissions average less than 10 per cent, representing the gross margin which has to cover all operating costs and yield a net profit. This sector of the intermediary market has also experienced a rapid growth in numbers. In the USA there has been an almost fivefold increase in agency outlets since the early 1970s. Of the 30,000 or more in existence in the early 1990s, more than two-thirds are single office agents and just a fifth branches of multiple firms. This is in contrast with Europe where single outlets number no more than a third of the total, the major multiples (100 or more branches) owning just over a quarter. In comparison with tour operators, the death rate record of travel agents is somewhat lower. Half of the agents currently in business have existed for ten years or more. In terms of sales, however, the multiples predominate. In the UK by the early 1990s, twenty agency firms had captured approaching half of the market, five alone controlling almost a third.

As in tour operations, economies of scale and scope characterize the travel agency sector but to exploit these fully in the UK, a company needs to own 150–200 branches. The principal economies are gained in the provision of central, mainly back-office, services such as accounting and computing. However, the experiences of multiples in the UK in the late 1980s do not verify that scale economies are actually achieved, where other smaller multiples were taken over. Because these struggling chains were not always profitable, the large multiples did not always increase their market share. What has become apparent is that the selective opening of new branches, sometimes in new forms such as in large stores or through franchising, has often been a better strategy than absorbing smaller enterprises (Liston, 1986).

Evidence provided by Liston indicates that travel agency, like tour operations in the UK, appears to be polarizing into the extremely large and the very small, although independent travel agents may co-operate to counter the dominance of the larger chains (Daneshkhu, 1997). The low entry costs contribute to explaining the large number of independently owned travel agencies but another important reason is that travel agency, as a retail activity, normally involves face-to-face contact with customers so that a more personal service is possible. Well-trained and skilled staff and close management control can be an asset to a business and engender a loyal clientele. The single independent travel agent or small local chain with a sound knowledge of the local market can perform better than the branch of a multiple, especially where specialization in certain services or types of tourism is required. In this sense, the large multiples with a uniform approach to selling holidays suffer diseconomies of scale.

With respect to spatial factors, an advantageous location is an important determinant of success, but this is offset by the costs of prime sites because of the competition to occupy these from not only other travel agents but also other forms of retailing. Nevertheless, spatial separation of travel agency lessens this competition and makes it possible for a firm to earn an adequate return on its investment. Though not conclusive, it partially explains the high proportion of independent agents in the USA, 60 per cent of which are located in suburban or small urban areas. Opinion is divided, both among academics (see for example McIntosh and Goeldner, 1990; Holloway, 1994) and practitioners as to the competitive structure of travel agents. Innovation, especially the introduction of information technology in the area of computer reservation systems, may prove to have a mixed effect (Bennett, 1993). Principals can potentially increase travel agency dependency and reduce competitors' impact by affecting the businesses that serve them, and the terms on which they do so, by utilizing a system specific to their own operations as the only means of communicating with the company, for example Thomson's TOPS (Thomson On-Line Program System). Such practices not only constitute a restraint on trade, so acting against consumers' interests, but also increase agency costs because of the need to set up the means of accessing a number of systems. They also inhibit the long-run achievement of economies of scale by the agencies.

The issue of whether trade associations perpetuate restrictive practices or promote greater efficiency in the intermediary sector is problematic. Those associated with travel agents have been able to secure as members a high proportion of those trading, largely because of the need to subscribe to schemes, such as bonds as a guard against the failure of tour operators, as a prerequisite for the procurement of a licence or to reassure customers. The American Society of Travel Agents (ASTA) and the Association of British Travel Agents (ABTA) are cases in point. Smaller businesses have also considered that these associations have given them a measure of protection, often by representing their interests to larger and powerful principals and public bodies. Of late, the function of these associations has been called in question as the cost of membership and protective schemes has risen sharply. Thus, many facets of the intermediary market are of current interest in economics, in particular, the study of company strategies and operations in what would appear to be both oligopolistic and competitive situations.

An oligopolistic structure prevails in the intermediary sector of the UK in that fewer than ten tour operators dominate the inclusive tour holiday market and determine the terms on which travel agents operate. This is particularly the case where tour operators have the potential to sell directly to the customer. Travel agents, therefore, are in the peculiar position of obtaining their inputs from an imperfectly competitive market but selling in one which is highly contestable. This peculiar structure again puts the theory of the firm to a severe test and the issue is one of the need for empirical investigation to establish the nature of the competitive structure of the travel agency sector, perhaps more in the context of an industrial economic analytical framework.

Transport

Given the wide range of transport modes, each possessing its own particular features and competitive characteristics and structure, transportation is best examined as a number of sub-markets. For example, the main commercial modes of air, bus/coach, ferries and rail suffer problems arising from indivisibility and associated fixed capacity and high fixed costs, periodicity and seasonality. None the less, there are considerable differences in their respective market structures and conditions. Their relative importance also differs in terms of the numbers of passengers carried, revenues generated and the degree of substitution possible. For international travel, air far outstrips bus/coach, sea and rail. Air travel has shown extremely high rates of growth since the 1960s, fed by technological change, with potential for continued growth in the future. Car ferry operations have benefited from the introduction of the roll-on roll-off vessels. Bus, coach and rail traffic have experienced relative decline in the face of the growth of private motoring.

The structures and competitive conditions of transport sub-markets are strongly influenced by the interrelationship between certain modes and the severity of the constraints imposed by regulation. There is also competition within modes, such as in the airline sector, where there is fierce competition on many short- and long-haul routes. Similar circumstances apply in the bus and coach market. In the ferry market, high volume routes tend to be subject to keen competition. There is a significant symbiotic relationship of modes. For instance, air travel depends on both bus/coach and rail for transfers to and from airports. Likewise sea crossings rely on road and rail transport for a large proportion of their traffic while the cruise-liner market is now largely based on air travel to and from the point of departure of the vessel. If the car hire sector is included, it is clear that it is dependent on airlines for much of its trade, often negotiating fly-drive packages as a means of increasing business. This complementarity, while it is an important feature of the transport market, does not suggest that there is not intense competition within and between modes. With respect to competition between modes, the opening of the Channel Tunnel linking the UK and France has demonstrated the

fierce competition of ferry and rail although the merging of some ferry operations raises the possibility of an oligopolistic market structure. In some
European countries, on domestic routes, air and rail travel are in competition,
especially where rail systems have been updated for high speed trains. Bus
and coach travel and rail also compete, the former offering a low cost alternative to the latter's shorter travel times.

Regulation of transport has occurred for two main reasons, safety and the
preservation of a market for a national or state-owned or supported mode.
Interest in the impact of regulation has centred more on deregulation with
the aim of fostering greater competition. However, it has become apparent,
as will be shown, that such an action often has unintended and opposite
effects, as has been manifested in air and bus/coach transport. Moreover, the
extent of regulation varies even within a specific mode. Again, in air transport, part of the market is more strictly controlled than others, creating considerable variations in competitive situations.

Air travel

The operation and regulation of air transport has been widely investigated
in the context of both business and holiday travel (Levine, 1987; S. Shaw,
1987; Button, 1991). Useful reviews of the situation in the early 1990s are
given in Ferguson and Ferguson (1994) and Lundberg *et al.* (1995) and
Melville (1995) has provided a thorough review and evaluation of economic
studies of the sector, including econometric modelling approaches.
Evidence of its conditions and structure illustrate a number of tourism and
economic issues arising in other transport markets, especially the economic
environment created by regulation and deregulation and the resulting level
of competition. An allied factor is the process of the privatization of a
number of airlines. The cost structure of airlines is quite complex and
operating costs and efficiency are related to technical factors relating to the
distance to be travelled, the size and type of aircraft and the payload, as well
as the support services such as marketing and the reservations system,
whereas capital costs are likely to be affected by whether aircraft are bought
outright or leased, new or secondhand. Direct and indirect costs, such as
airport charges and ground handling, are outside the control of the airline.
Prima facie, airlines are high fixed cost enterprises with fixed capacity so
that they need to attain high payloads to break even. Short-haul flights,
involving fewer hours in the air and more take-offs and landings, are relatively more expensive than long-haul carriage. Furthermore, aircraft are
designed to fly specific distances as efficiently as possible so, for example,
employing a Jumbo jet on European routes of up to 2,500 km is uneconomic. In general there are economies of scale, stemming from technical
efficiencies in operating large aircraft provided the routes operated on warrant it. In addition they depreciate less quickly. Uncertainty is a significant

79

factor because of the rate of technical development of aircraft, for example with respect to aerodynamics and engine efficiency.

The market served influences the economics of airline operations in a number of ways. For scheduled services, to maximize the payload and revenues, demand has to be ascertained in advance to discover its composition with regard to the numbers of business and economy customers, for example. The approach being adopted by airlines is termed 'yield management' which in its most developed form is akin to perfectly discriminating monopoly pricing. Pricing policy also needs to take account of the route, destination and stage stops, the pressure of competition and longer-term level of demand. Chartered services reduce market uncertainty because it is possible to ascertain the payload well in advance. However, the distinction between these two main types of services has become more difficult to discern, as charter airlines have moved into the scheduled market and vice versa. This has occurred because travel on what were holiday charter routes has become so regular as to warrant scheduled services. Conversely, some under-used scheduled services, perhaps operated as a social service when airlines were state-owned, may be served more irregularly by special charters.

The advent of deregulation and privatization has dramatically changed airline operations and has had some interesting economic consequences. The US experience, where deregulation occurred in 1978, holds implications for what might happen elsewhere, such as in the European Union where moves to relax controls have been initiated. Competition has certainly intensified as new companies moved into the market, entry having been made easier. This tends to suggest that entry costs, particularly the capital costs, are lower than casual observation indicates. Chartering, leasing or buying aircraft second-hand facilitates ease of entry, as does negotiating reservation, ground handling services and maintenance agreements with specialist contract firms or established and larger airlines. In the short run the increased competition has brought fares down and extended the market, also forcing airlines to be more sensitive to their customers' needs.

However, recent developments have thrown up the negative effects of deregulation with some former major airlines having such severe financial problems as to be forced out of business, Pan Am in the USA being a well-known example. Also some major national airlines in Europe, such as Air France, Iberia, Sabena and SAS, have had problems in adapting to severe competition, especially on key routes, mainly because of inflexible structures, engendered by highly restrictive international fare policies and state support as national flag carriers. The benefits to consumers may be short-lived as the number of airlines will almost certainly decrease in the long run through mergers and reciprocal agreements and the failure of smaller, newer entrants. This concentration and consolidation is likely to lead to higher fares as the larger companies take control of key routes and gain the market power to dictate price levels and limit choice. The development of the hub and spoke pattern of air travel where long-haul flights are concentrated on major airports gives opportunities for smaller feeder airlines to create a market niche, although the larger airlines may determine the terms on which spoke routes will be served. There are additional concerns where airlines cut costs in order to remain competitive as the quality of the service may be lowered, less attention being paid to maintenance and safety margins.

The pattern of air travel is evolving given the differences in the rate of growth in sub-markets and the fact that deregulation has not yet had a full effect on the airline structure and the market. However, there are signs that market power is likely to be concentrated in a few international giants, especially as state support for national airlines declines under privatization programmes and agreements, such as that in the European Union, which generally forbids subsidization of unprofitable airlines. A case in point was the proposed agreement between British Airways and American Airlines, two of the largest airlines in the world, which control around 60 per cent of the UK–US routes.

The evidence from this brief description of what is the best researched tourism sector is that although a domestic monopoly or oligopoly structure has been common, with a single state-supported airline or a small number of competing airlines, deregulation has made some markets competitive in the short run. In the international market some routes are competitive, being served by many carriers. Most of the others are served by at least two carriers, indicating an oligopolistic market, although a few routes are served by a single carrier which may be tempted to exercise monopoly powers. This does not undermine the ability of the theory of the firm to explain the market structure; it signifies a need to divide the sector into sub-markets and to consider each separately. The sector, in common with the others reviewed, is in a fluid state, reflecting deregulation and the changing demand for foreign air travel.

Other transport sectors

The structures of the bus, coach, ferry and rail sectors are similar to that of air travel in that they, too, experience the problems of high capital costs, fixed capacity, peaked demand, the need for feeder routes to sustain profitable ones. Some state support and regulation characterize these modes. On the other hand, there are opportunities to exploit economies of scale and exercise price discrimination to maximize revenues and fill any excess capacity. For example, the principal problem of the rail sector is the high fixed costs associated with the track, signalling system and stations. In the UK, the privatization process has been facilitated by separating this basic infrastructure from the rolling stock, making it possible to reduce entry costs and to perceive it as resembling air travel where airports and services, navigational and air traffic control systems are operated by bodies on behalf of the airlines.

The crucial difficulty in deregulating and privatizing rail markets and operating them commercially is that uneconomic lines may be closed, although they have the potential to support the principal inter-city services on the same basis as the evolving hub and spoke structure in air travel. Another similarity with air travel is the endeavour to fill excess capacity by a complex pricing system resembling a crude form of yield management.

A number of elements of the bus and coach market resemble the air and rail sectors. The main issues in the market in the mid-1990s are the standard of service, safety and reliability, with deregulation in the UK creating uncertainty for both long-established and new entrant operators. The aim of deregulation, as in the air travel market, is to engender greater competition. A large number of new operators has given rise to substantial overcapacity. Although this initially lowered prices on key routes, the entry of many small companies has been accompanied by greater concentration at the national and regional level of operations. Large companies not only offer a more comprehensive coverage but also are able to exploit economies of scale in such factors as reservations, repairs and maintenance, administration and advertising, as well as more fully utilizing vehicles with respect to payloads and operational hours and operating only the most lucrative routes at higher fares, endangering the long-run existence of the smaller companies. Consequently, market power may be vested in few hands.

The ferry market clearly serves the private road, coach and rail transport modes and has grown strongly with the increase in surface holiday travel, particularly in Europe. New forms of conveyance, such as hovercraft, hydrofoil and catamaran craft, have reduced travel times and extended the market, increasing both intra- and inter-mode competition. Short routes involve high entry and operating costs, somewhat resembling short-haul air travel. There are economies of scale in increased vessel size and speed but these can be offset by the danger of increasing loading and unloading turn-round times and excess capacity in off-peak and off-season periods, akin to those in the air, bus, coach and rail market, giving rise to a complex fare structure and travel conditions. Companies rely heavily on on-board spending to bolster revenue. On longer-haul routes the problems of periodicity and seasonality are more acute and where remote island communities are served, state subsidies are required.

The existence of the cruise-liner sector as a sea-based form of tourism needs to be acknowledged if only because it is a growth sector, especially in the USA, which accounts for over three-quarters of the global market. It is not considered here except to note that from an economic standpoint it has many of the same characteristics as air travel, being capital-intensive with high fixed costs for operators who own rather than charter ships and depending on high volume and repeat bookings to fill the fixed capacity. Technically there is a parallel between the type of vessel and aircraft which can operate at optimum efficiency so high payloads are essential.

KEY FEATURES OF TOURISM SUPPLY

Further analysis of the structure of tourism supply and the degree and type of competition between firms, described in the preceding overviews of tourism sectors, is facilitated by reference to the theoretical models presented at the beginning of the chapter. The models identified a number of criteria which indicate the competitive structure of markets; for example, the number and size of firms and level of entry/exit barriers indicate the degree to which oligopoly or monopoly powers may be exercised by tourism firms. Similarly, the extent of market concentration or of price leadership signifies potential limitations on the level of inter-firm competition. These criteria can be examined in the context of specific tourism sectors in different countries, providing insights into the types of market structure which prevail and the nature of inter-firm competition. This is useful because of the implications of different market structures for the extraction of profits above the break-even level and, hence, for consumer welfare.

Key criteria which were identified within both the theoretical models and the overviews of tourism sectors are

- number and size of firms
- degree of market concentration and level of entry/exit barriers
- economies and diseconomies of scale and economies of scope
- capital indivisibilities, fixed capacity and associated fixed costs of operations
- price discrimination and product differentiation
- pricing policies – leadership, wars and market-share strategies.

The first four indicate market structure, while the last two also relate to the strategies which firms pursue within imperfectly competitive markets. They will now be discussed in more detail, linking the theoretical and empirical discussions of this chapter, as well as acting as a bridge to the more industrial economics-oriented analysis of Chapter 5.

Number and size of firms

Where many small firms exist, the inference is that the market is competitive. A small number of firms, in contrast, indicates an oligopolistic structure, with monopoly being the limiting case of dominance by a single firm. Evidence on the numbers and sizes of firms in the transport and intermediary sectors is relatively easily ascertained where there are requirements regarding registration and regulation and trade associations exist. For example, the International Air Transport Association (IATA) and International Civil Aviation Organization (ICAO) publish statistics on airlines while in the UK and many other countries, there are intermediary trade associations representing tour operators and travel agents which allow estimates of the numbers of intermediaries to be made. However, the size, diversity and fragmentation of the accommodation sector mean that except for hotels of an international level, ascertaining the number of establishments is extremely difficult.

The size of firms can be measured by a number of variables, ranging from number of employees, sales revenue, number of unit sales to capital employed. In the accommodation sector, the number of rooms offered by each establishment or the total number of bednights achieved are useful measures. The number of holidays sold is a feasible measure for intermediaries. Examples relating to transport are the number of passengers carried or passenger kilometres which can be compared for different carriers and the proportion of the total market or segment served. In the case of air transport, IATA statistics reveal that in the early 1990s, for international travel, British Airways carried by far the largest number of passengers and had the greatest passenger kilometres, followed by Lufthansa, Air France and Japan Airways. In the UK intermediary sector, Civil Aviation Authority (CAA) data showed that in the mid-1990s, the largest tour operator in the air inclusive tour (AIT) market was Thomson, accounting for nearly a quarter of the total number of holidays sold.

The composition of tourism sectors within and between countries differs in terms of the number and size of firms. In some countries, a domestic monopoly exists in the airline sector, as in the case of Aloha Airlines in the Hawaiian islands. In others, an oligopolistic structure prevails; for example, in Australasia, the air transport sector has been dominated by two to three airlines operating reciprocal agreements. Within the accommodation sector, the large number of establishments selling differentiated products is suggestive of monopolistic competition, while some bus and coach sectors have been highly competitive. However, it is also clear that although in some sectors a few large firms command a substantial market share, the remainder of the market consists of many highly competitive small firms. Hence, it is possible to conceive of a spectrum of competition with strongly competitive conditions coexisting with oligopoly in some tourism markets, with a degree of interaction between them. For example, innovations by small companies may subsequently be taken up by larger entities, as in the case of the timeshare concept or activity holidays. Thus, tourism markets are complex and may consist of different sub-sectors characterized by alternative competitive structures.

Degree of market concentration and level of entry/exit barriers

The degree of concentration is a further indicator of the competitiveness of markets. High concentration is suggestive of oligopoly or monopoly while a low incidence implies a high level of competition. Measures of concentration can be broadly divided into those which focus on the number and size of firms and those which consider the implications of variations in size. Size may be defined in several ways, as mentioned above, the indicator used depending partly on the availability of data and partly on the nature of the business. The Lorenz curve can be used to rank, in cumulative form, the smallest to the largest firms as a percentage of the total number of firms trading in the market. The Gini coefficient represents the magnitude of the divergence from exact equality in size of firms, so that a value of zero denotes that all firms are of equal size and, ostensibly, highly competitive conditions prevail, whereas a value of unity equates to pure monopoly. The concentration ratio ranks firms from largest to smallest, so that a high percentage score indicates a market in which monopoly prevails, while a low score suggests that many firms serve the market. More sophisticated measures attempt to take account of the greater impact of larger firms on the market by weighting their market share more heavily. Well-known examples of such measures, widely applied in the industrial economics field, are the indices proposed by Herfindahl and Hirschman (Hirschman, 1964) and Hannah and Kay (1977), evaluated by Davies (1989a). The degree of market power can be measured by the Lerner index (Lerner, 1934) which considers the discrepancy between market price and marginal cost, where zero denotes equality and a highly competitive market and a value approaching unity indicates high profits and a highly concentrated market.

The degree of market concentration in different tourism sectors varies between sectors, countries and/or regions and key tourist locations of a given country. In Spain, the five main charter airlines were responsible for over 80 per cent of all passenger kilometres flown by Spanish charter airlines at the end of the 1980s while in the UK, the top five firms carried approximately 70 per cent of all charter passengers (Bote Gómez and Sinclair, 1991). At the same time, the top five UK travel agents had a market share of 47 per cent compared with a share of 77 per cent for the top five tour operators.

The extent to which a higher degree of market concentration is accompanied by profits in excess of the break-even level is determined, in part, by ease of entry into and exit from the sector and the level of sunk costs involved in operating the firm. Traditionally, attention has concentrated on 'innocent' entry and exit barriers determined primarily by technological requirements, principally the scale at which firms can operate efficiently. However, firms may also engage in strategic entry deterrence in an attempt to maintain relatively high prices and supernormal profits over the long run. Entry and exit conditions and other forms of strategic behaviour by firms will be considered in Chapter 5.

Economies and diseconomies of scale and economies of scope

In order to relate economies and diseconomies of scale to models of market structure, it is necessary to look again at economic theory. The essence of economies of scale is that supply costs per unit of production decline as inputs are increased and output expands. Clearly, as long as average unit costs are falling, there is an incentive for a firm to go on enlarging. The point where unit costs start to rise again, i.e. the lowest point on the average cost curve, is where diseconomies set in and the firm would no longer expand. In economic theory the issue is whether diseconomies are reached at a relatively small or large size of firm. The former implies many small firms operating in a highly competitive market, trading at prices which equate with lowest average cost, while the latter suggests the converse.

Thus, in conjunction with the number of firms, their size and the level of barriers to entry, evidence of the extent of economies or diseconomies of scale can indicate market structure (Lyons, 1989). In order to establish whether either exist in tourism, it is necessary to look at the cost structures of firms of differing sizes. In some markets, as shown earlier in relation to the transport sector, particularly air and sea travel, there are technical reasons which explain why larger aircraft and ships are more efficient, in terms of both operation and maintenance, therefore reducing per unit passenger costs. Furthermore, there are economies of scale in production through specialization. Increasing size in the accommodation sector also yields some economies of scale with respect to the management, staffing and servicing of a greater number of rooms and in the travel agency sector overhead costs can be spread over additional outlets.

Economies of scope occur where additional products share common inputs. For example, in advertising and marketing new products, suppliers can draw on existing resources. Thus, airlines, bus, coach and ferry companies may add new routes, more fully utilizing the means of conveyance. Travel agents and tour operators can serve new market segments using existing reservation systems and staff, while in the accommodation sector, the market may be extended via the supply of self-catering in addition to serviced holidays or access to the establishment's facilities by non-residents. To an extent, therefore, opportunities for economies of scope are possible where there are capital indivisibilities or spare capacity. These are considered in the fourth feature of tourism supply examined below.

Evidence of unexploited economies of scale or scope, or potential for them to occur in the future through technical change, would indicate the likelihood of increased firm size and greater market power, so rendering the market less competitive. Conversely, diseconomies of scale, or supply where widely differing input proportions are possible, would enhance the long-term viability of smaller firms and create a structure which is more competitive. The discussion of tourism supply sectors indicated that although large firms can incur economies of scale and dominate some markets, this does not preclude many small firms from being successful in the same market. The latter can often serve specialized segments which are not susceptible to large-scale operations and some activities remain small scale because the markets are local or national rather than international; enlargement of their activities would generate diseconomies of scale. In general, the principal transport sectors are characterized by the possibilities of economies of scope. It is in the accommodation and intermediary sectors where greater polarization has occurred.

Capital indivisibilities, fixed capacity and associated fixed costs of operations

Many tourism sectors utilize capital equipment which is indivisible, aircraft being obvious examples. The firm often possesses a set amount of equipment and thus fixed capacity over the short run and incurs fixed costs in maintaining it; it can be altered only in the long run when re-capitalization occurs. As capital is used, variable costs arise, for example, in the form of fuel and staff costs in the case of air transport. In the short run, economics argues that a firm, even if it does not cover its total costs, should continue to trade as long as it covers its variable costs. However, in the long run the firm must generate sufficient revenue to cover both its fixed and variable costs if it is to remain in business. There may also be sunk costs which are not recoverable if the firm leaves the sector. They normally relate to indivisible inputs and can affect the contestability of the market as they deter entry into a sector if wholly irrecoverable. For example, if an airline, bus or ferry operator finds a route uneconomic or a hotelier in a particular location finds demand to be lower than the break-even level, each would, ideally, wish to execute a costless withdrawal. This depends on whether assets can easily be redeployed or, through marketing and pricing strategies, market demand increased to fill space capacity.

It is useful to analyse the structure of costs to ascertain both the absolute level of costs and the proportion which is fixed, as opposed to variable. In the airline sector in the early 1990s, International Civil Aviation Organization statistics indicated that fixed costs were high in absolute value and were of the order of two-thirds of the cost of running an airline, if skilled aircrew are taken as a fixed input and aircraft need to be stored and maintained if not operational. The variable costs are those which would be avoided if the aircraft were withdrawn from service, mainly fuel and landing charges and some on-board consumables and aircrew costs. They account for around one-fifth of the total costs although there is obviously some variability between airlines and routes. For rail systems, the infrastructure and rolling stock create high fixed costs and relatively low variable costs. In absolute terms, bus and coach services are not characterized by high fixed costs but in relative terms, particularly for the smaller operator, they are significant.

In the accommodation sector, the fixed costs as a proportion of total costs are much lower than in the transport field (Horwath Consulting, 1994) and the value of fixed costs associated with small accommodation establishments are often low. Capital and recurrent costs related to hotels in the UK account for about one-third of total costs. Overall, with items such as insurance and other capital inputs included, fixed costs are between half and three-fifths of total costs in the UK. In the intermediary sector conditions are rather different, depending upon whether tour operators and travel agents operate their own aircraft and accommodation. In the case of small operators who are not vertically integrated, fixed costs are usually low in relative and absolute terms. Evidence of high fixed costs coupled with fixed capacity in some sectors is indicative of the emergence of large firms which can exploit economies of scale, increasing the degree of concentration.

Operations subject to high fixed capacity and fixed costs require relatively high occupancy rates or load factors in order to meet both fixed and variable costs; the distinction between profits and losses occurs at occupancy rates around such break-even points. Cyclical and seasonal variations in tourism demand exacerbate profits volatility, especially where aggregate fixed capacity is a feature of independent decisions made by many suppliers such as in the accommodation sector. Strategies in the short run are aimed primarily at covering fixed costs, involving attempts to capture trade from rivals. In the long run, consideration is given to means of raising demand in order to increase load factors or occupancy rates, particularly in off-peak periods. The characteristic of high sensitivity of profits to occupancy rates and load factors engenders various responses by suppliers. The strategies which they adopt concern pricing, the product, consumers targeted and markets or segments served are analysed in Chapter 5.

Price discrimination and product differentiation

The pricing and output policies practised by tourism firms provide some indication of market structure. A large number of firms producing relatively homogeneous products which must be sold at the prevailing price in highly competitive markets approximate to the model of perfect competition. Taxi services in many tourist destinations are a case in point. In imperfectly competitive situations where firms have a degree of control over the market, pricing and output strategies can be implemented. There is considerable interdependence in the pricing and output policies of oligopolists as they have to take account of their rivals' actions when determining their own strategy. In order to extend their share of the market, firms operating under monopolistic competition endeavour to differentiate their products from those of other firms. However, given the appropriate conditions (discussed below), firms operating under all types of imperfect competition can benefit from price discrimination, which involves targeting specific groups of consumers. Thus, its existence indicates an imperfectly competitive market.

Price discrimination is based on the idea that many consumers, who purchase goods and services in markets where a single price prevails, enjoy a welfare gain known as consumers' surplus because they would have been willing to buy at higher prices, sometimes referred to as their reservation price. Suppliers who recognize this can discriminate between purchasers and charge higher prices to those with higher reservation prices, and vice versa. There are various degrees of discrimination depending on how many groups of customers can be identified. In the extreme case it is hypothetically possible to exercise perfect, or first degree discrimination where every consumer pays a different price. Firms gain increased profits provided that the price at which additional units are sold is greater than the cost.

The most likely situation in which price discrimination can be employed is where different consumers have distinct price elasticities of demand, permitting the supplier to charge higher prices to those whose demand is inelastic while others, whose demand is more elastic, are charged lower prices. The supplier wishes to divide the total units of the product available between the different groups of consumers so that marginal revenues for each group are equal. This determines the price each group should be charged, enabling the firm to maximize profits, and is illustrated in Figure 4.7.

Consumers with an inelastic demand change their demand for the product by only a small proportion in response to a proportionate change in price and their demand is shown by the average revenue curve, $AR_1 = D_1$ in the figure. The corresponding additional revenue which the firm receives for each extra unit of the product sold is shown by the marginal revenue curve, MR_1. The average and marginal revenue curves for a consumer with more elastic demand are flatter and are given by $AR_2 = D_2$ and MR_2. The marginal revenue

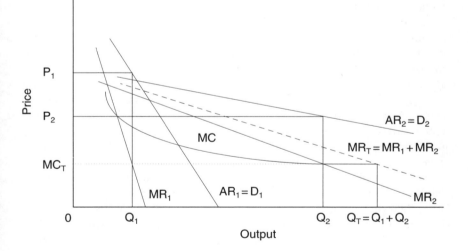

Figure 4.7 Price discrimination between consumers with different price elasticities of demand
Note: P = price, Q = quantity of output, MC = marginal cost, MR = marginal revenue, AR = average revenue, D = demand

curve for all consumers is indicated by the broken line as $MR_T = MR_1 + MR_2$ and is obtained by adding horizontally the marginal revenue gained from each group of consumers. The marginal cost curve, MC, shows the additional cost incurred by the firm for producing each extra unit of production. The total profit-maximizing output occurs where $MC = MR_T$, giving rise to output $Q_T = Q_1 + Q_2$ and a marginal cost MC_T. The firm divides its total output between the two groups of consumers, selling a quantity Q_1, determined where $MR_1 = MC_T$, at a relatively high price, P_1, to consumers with an inelastic demand. A higher quantity, Q_2, determined where $MR_2 = MC_T$, is sold at a lower price, P_2, to consumers with a relatively elastic demand.

There are a number of examples of discriminatory pricing in tourism. It is possible to observe price discrimination in air, ferry and train travel and in the accommodation sector. For example, airlines perceive their passengers as falling into three broad groups – first class, club and economy – in which first and club class seats are filled mostly by business travellers. In ferry operations, commercial demand is generally more price inelastic than holiday demand. For rail travel, the division between commuting and pleasure demand is particularly significant, so that low prices are offered at times of low weekday and weekend demand in an attempt to attract travellers with a relatively high price elasticity of demand, while business people who often need to travel in peak periods are charged relatively high prices. In the accommodation sector, the larger hotels which have spare capacity offer weekend bargain breaks, charging guests with inelastic demand higher prices during the working week. Airlines, bus, coach and ferry companies also follow this practice in the off-season.

Market segmentation and product differentiation are also common practices under imperfect competition. Product differentiation can take the form of vertical differentiation between different quality products so that, for example, some tour operators attempt to specialize in providing luxury holidays in expensive locations. It may also involve horizontal differentiation via the supply of a range of product types, such as the provision of holidays for mass market demand as well as for the upper segment of the market, young people, elderly people and a wide range of special interest groups. The strategy of branding aims to raise consumer awareness of and demand for particular product types and advertising performs a similar function.

Few smaller firms have either the capacity or the market power to exercise significant control over prices so that serving a specific segment or differentiating their product is the only option open to them. Their concern is often to create rather than extend a market. Effectively they look for an unfilled niche and attempt to meet a specific demand in what may be a relatively small market segment. To this extent the product is automatically differentiated. It is large firms, enjoying economies of scale and scope and spare capacity, which attempt to differentiate their products through branding in addition to engaging in price discrimination. In the accommodation sector many international hotels serve the top end of the market, very often for business travellers, while budget hotels, for example Travelodge and Budget, cater for families. In some cases, such as the Accor group, both markets are targeted under different brand names.

Pricing policies – leadership, wars and market-share strategies

Pricing strategies are central to oligopoly and may be used in attempts to maintain or increase market share and/or to eliminate excess capacity. The strategies of Laker Air in the 1970s and Virgin in the 1990s are prime examples and the advent of deregulation has engendered further price competition among airlines on key routes. As the economics of oligopoly predicts, price-cutting by one airline provokes a response from others. This occurred in the case of Trans World Airline (TWA) when it cut fares within the United States in 1991. The matching of these fare cuts by rivals with a lower operating cost base partially explains the airline's bankruptcy. The air transport sector also contains small airlines which are relatively efficient over particular routes. Thus, British West Indian Airlines (BWIA) is cost-effective on a number of Caribbean routes (Melville, 1995) and British Midland is able to compete on domestic and European inter-city routes despite the domination of the market by British Airways.

Price-cutting has also occurred within other transport sectors, such as bus, coach and ferry services. In the first it is very much the outcome of deregulation, while on heavily used routes in the last, although there is intra-sectoral competition between companies, the price 'war' has been driven more by inter-sectoral competition. A feature of some interest is that many air, bus and coach sectors have experienced significant volatility because of structural changes which have destabilized trading conditions and prices. Ferry services are in a slightly different position because the entry of new companies is uncommon and there has often been tacit or overt collusion between the main ferry companies, giving reasonable stability in fares. This was also the case prior to deregulation in air travel where air fares were more formally controlled by IATA, resulting in a virtual cartel. In the accommodation sector, price wars have been less evident because there is greater separation of establishments in the market in terms of type, quality and location. Where establishments are clustered, such as the larger hotels in holiday resorts, near airports or in city centres, there may be implicit price-fixing because operators have easy access to information on competitors' charges and can set their own prices accordingly.

The almost classic demonstration of the oligopoly case in tourism supply is package market tour operations. Here price leadership and wars and attempts to increase market share by the larger players are strategies which have been adopted since the advent of mass tourism in the 1960s. The evidence for the UK, including an historical review, can be found in N. Evans and Stabler (1995), following the studies of competitive structure and strategies by Sheldon (1986) and Fitch (1987). Evans and Stabler suggest that structural changes are occurring, such as the entrance of corporate tour operators, often part of a conglomerate, which emphasize the gaining of profits rather than market share. There are also discernible trends in demand towards more specialized holidays. The mass package market is dominated by the top tier of approximately ten tour operators, who have increasingly diversified into specialized types of holidays to exploit economies of scope. At the other end of the firm size spectrum are small specialist operators which can survive through niche marketing, although there has been a high birth and death rate of companies. It is the second tier of thirty or so firms which has lost part of its share of the mass market to the top ten operators and the more specialized segments to the smaller firms.

Price leadership and pursuit of increased market share is mainly undertaken by large groups through longer-term strategies such as mergers or takeovers. Increased integration between firms appears to be a logical outcome of tourism markets where there is a long chain of supply. Large tour operators in the UK are vertically integrated, particularly with travel agencies and airlines, while in the USA there are strong linkages between travel agencies, the accommodation sector and transport modes, with the initiator tending to be the transport company. Reciprocal agreements, franchising and leasing constitute forms of integration other than ownership ties. The relevance of integration to competitive structure is that its existence suggests information advantages, cost savings and higher profits effected by enlargement of the firm and the possibilities of increased market power, associated with oligopoly and even duopoly or monopoly. At an international level, integration has ramifications much wider than those appertaining to the suppliers and tourists themselves, particularly in developing countries, as discussed in Chapter 6.

CONCLUSIONS

The outlines of the structure of the accommodation, intermediary and transport sectors and the discussions of their key features reveal that market structures can be quite heterogeneous within each sector. They possess such characteristics as a wide range of competitive forms, market segmentation, product differentiation, high rates of entry and exit, some scale economies and significant variations in the degree of regulation. Moreover, there appear to be considerable changes in the structure of tourism supply in terms of the number of firms, their size and market share, particularly in the intermediary sector. These characteristics raise two questions. First, how far does the neoclassical theory of the firm and the competitive structures it has identified explain what is observed in tourism supply? Second, does heterogeneity within a specific sector suggest that the theory of the firm is inadequate or does it merely signify that the categorization of tourism sectors and markets adopted currently is too broad?

With regard to the first question, the test of the appropriateness of the theoretical concepts is whether they can explain actual market structures and predict the outcome of changes occurring in the industry or its markets. It appears that in each of the three main sectors considered, there are elements of contestability (whereby new or established firms can engage in price competition with existing firms) alongside the dominant market forms of monopolistic competition and oligopoly. This contestability aspect is not easily encompassed within the conventional theory of the firm because its static equilibrium analytical framework is not well suited to accommodating what is effectively a dynamic situation. Thus, while it is possible to explain the conditions of tourism supply sectors as being indicative of particular theoretical competitive structures, it is more difficult to predict the outcome of changes in the conditions. For example, in the transport sector the possibilities of exploiting economies of scale should give rise to increasing concentration and few firms, therefore denoting an oligopolistic structure, as appears to be the case in the air, bus, coach and ferry sectors. Furthermore, the theory would indicate reasonably stable prices, given the constraints which individual firms face in raising or lowering them because of possible reactions by rivals. However, deregulation has effectively reduced entry barriers which has an impact running counter to that of economies of scale. It has created greater volatility in prices and led to an influx of smaller companies and greater competition in the short run. Although, after initially successful trading, there are signs of a decline in the number of firms because of failures by new entrants, particularly in the bus and coach sector, it is not certain that this will occur in the airline sector. Thus, the predictive power of conventional theory is called into question.

In the intermediary sector, particularly tour operations, increasing concentration is occurring in what conventional theory would posit to be a highly competitive market because of low entry–exit barriers and limited economies of scale. Paradoxically, the largest tour operators in what might be termed the oligopolistic segment of the UK holiday market have been engaged in fierce price competition and, over many years, margins and profits have been low or even losses sustained. Conversely, the many smaller specialized tour operators have avoided such competitive conditions. A similar scenario applies in the UK travel agency sector. The long-running pursuit of increased market share in the intermediary sector appears to be at odds with theory which assumes profit-maximizing behaviour.

The second question as to the observed competitive heterogeneity of most tourism supply sectors and the relation to the theory of the firm is problematic. Analysis within the conventional framework tends to compartmentalize markets into clearly defined competitive structures, seldom acknowledging that sub-markets may be subject to different conditions. Does this matter? Is the theory undermined by recognition that in particular markets different competitive forms coexist? The response by mainstream economists might be that the various segments should be treated as separate markets. However, this may not be appropriate for it has already been demonstrated that there is an interrelationship between the differing competitive segments within a single sector. For instance the largest firms within the transport sector do react to the behaviour of smaller and newer rivals. Accordingly, the further sub-division of tourism markets to meet traditional analytical requirements may not be appropriate.

While acknowledging that a full exploration of the applicability of the theory of the firm has not been undertaken, the illustrative examples indicate the limitations of its explanatory and predictive power in the context of tourism supply. The theory aids identification of the key variables determining individual firms' and industries' behaviour within each of the competitive forms and it acts as a rigorous foundation on which to develop more complex and differentiated models. Increasingly, the economic analysis of industry is recognizing the role of institutional structures and their possible evolution, together with the need to examine the dynamic nature of markets where information availability uncertainty and transaction costs are significant. Attention, therefore, is now turned to a more industrial economics-oriented perspective where a number of features of tourism supply, perceived as important to its structure and operation, are identified and examined.

5

THE STRUCTURE, PERFORMANCE AND STRATEGIES OF TOURISM FIRMS

INTRODUCTION

The neoclassical analysis of market structure outlined in the previous chapter proved useful for identifying different types of markets and giving a number of valuable insights into aspects of firms' conduct and performance. However, such phenomena as imperfect competition, particularly oligopoly, and uncertainty, as well as the need to explain market dynamics, are inadequately explained by traditional analysis. The development of industrial economics and other schools of thought have attempted to fill the gaps left by traditional approaches and illustrate the theme of this book not only by indicating recent theoretical developments and advanced economic analysis but also by helping to explain tourism supply. Within industrial economics, two main approaches have been used to examine the conduct and performance of firms in different types of markets. The first is the structure, conduct and performance (SCP) paradigm which has played a major role within empirically oriented studies of firms within manufacturing industries. Although it has been subject to some criticism, particularly in more theoretically based analysis, the paradigm remains a useful framework and appears to be relevant to a complex service industry such as tourism. The second, more recent approach is game theory which is used to analyse the strategies which firms adopt in relation to the actions and probable reactions of their competitors. Game theory has been applied extensively in oligopolistic situations and has vastly improved the comprehension of firms' interactions and their outcomes in dynamic situations. It seems particularly important for understanding the behaviour and strategies of tourism suppliers.

Before considering these analytical developments, it is instructive to review briefly the approaches which have been adopted in economics with regard to central issues concerning the market environment in which firms exist and the ways in which it can influence their structure and behaviour. First, the main schools of thought which are relevant to industrial economics and which can assist the explanation, as opposed to the description, of tourism supply and changes in it are discussed. Therefore, elements of the Austrian school and behavioural, evolutionary, institutional and psychological economics are outlined. Although they are not applied to tourism supply owing to constraints of space and the absence of tourism literature on the subject, they are useful in providing the context for the ensuing discussion and a basis for future research. The SCP paradigm is then explained and its application to services, using the case of tourism intermediaries, is indicated. The contribution and limitations of the approach will be evaluated. The role which game theory can play in explaining the strategies of tourism firms in a dynamic context and the associated changes in the structure of tourism markets will be examined subsequently. An overall assessment of the two approaches is provided in the conclusion.

THE INDUSTRIAL ECONOMICS BACKGROUND

The SCP paradigm (Chamberlin, 1933; Bain, 1956; Mason, 1957) dominated industrial economics until the 1980s. According to the paradigm, it is the type of market structure within which firms operate which is the ultimate determinant of their conduct and performance, measured by such criteria as profitability. Market structure variables such as the number of buyers and sellers and degree of market concentration are assumed to be relatively stable. As an empirically based analytical framework, the SCP approach accepts that market structures usually differ from the benchmark of perfect competition, so that if firms set prices in excess of marginal cost, there is a prima facie case for government intervention and the implementation of measures to bring about increased competition. The SCP approach is, thus, policy-oriented.

In contrast to the more empirically oriented SCP model, the Chicago school continued to adhere to and propagate an approach based on the conventional neoclassical model of long-run equilibrium within competitive markets. Its thesis is that resources are allocated optimally by competitive forces within the market and that market power and pricing in excess of marginal cost are eliminated by free entry into the industry. A notable contribution, partially modifying the somewhat restrictive conventional position, was that of Baumol (1982) who, reflecting concern that traditional models failed to explain competitive structures fully, promoted the theory of contestable markets, discussed in Chapter 4. The Chicago school views perfectly competitive, or at least contestable, markets as the dominant market structure which emerges over the long run so that consumers benefit and government intervention is regarded as unnecessary. The Chicago stance was adopted by some economists in Europe, as indicated in the reviews by de Jong and Shepherd (1986) and Hay and Morris (1991). There has been a long-running controversy between the Chicago economists and proponents of the SCP paradigm, broadly supported by the behavioural, evolutionary and institutional schools of thought, who do not adhere to the view that markets are generally competitive over the long run (S. Martin, 1993).

Research on the economics of industry undertaken outside mainstream analysis has moved the debate on the competitive structure of the market into a different arena. There are two distinct but related strands to this work. One has concerned the accommodation of the dynamic nature of markets while the other has concentrated on the representation of the characteristics and circumstances of firms, industries and markets. The Austrian school, associated with scholars such as Hayek, Menger, Mises and, initially, Schumpeter (disavowed by neo-Austrians), is notable for its concentration on the process of competition as opposed to the static equilibrium analysis embodied in much conventional market structure analysis. These scholars acknowledge that change and uncertainty are endemic and that those involved in industry have to make decisions within this context. However, they argue that over time, with experience and the benefit of increased knowledge of processes and opportunities, better decisions are made in succeeding periods, thus tending to produce more competitive market conditions. In this sense there is a degree of what Hayek (1949) calls emerging 'order' but not necessarily equilibrium. Latterday researchers who can be viewed as constituting the neo-Austrian school are Kirzner (1973), Reekie (1984) and Littlechild (1986). This view of industrial activity is paralleled by what R.R. Nelson and Winter (1982) refer to as 'Evolutionary Economics' in which procedures for carrying out required actions evolve with the accumulation of knowledge.

The development of evolutionary economics has run in parallel with that of institutional economics and, in part, the thinking of the Austrian school. It has been concerned primarily with treating endogenously the way the beliefs, norms and customs of society might lead to the formation of institutions which facilitate corporate behaviour. Theories of the evolution of institutions are based on those employed in biology and mathematics involving natural selection. Hirshleifer (1982) reviewed these when modelling social and institutional change, arguing that game theory can indicate the outcome of co-operative and non-cooperative behaviour to show the desired human traits which would be required to devise effective institutions. The relevance of these notions to economic organizations such as the firm is to underpin the importance of the attitudes and objectives of entrepreneurs, managers and workers in determining the operation of industries and markets. An indication of this emerges in the examples of non-cooperative games presented towards the end of this chapter. Tourism supply is an interesting case because it displays, through the instability in some sectors, a measure of immaturity with respect to the evolution of its business organizations so that outcomes are often less predictable than in mature and stable markets, also reflecting a possible distinction between manufacturing and service sectors.

In emphasizing the dynamic nature of the market system and its institutions, the Austrian and Evolutionary Economics schools echo Marx (1967). In his analysis of competition, Marx was concerned not only to reveal its process but also to ascertain the outcome in terms of the impact on the distribution of income and wealth and the allocation of resources. Followers of Marx have tended to underplay his perception of competition and accentuate his view that industrial production would become more concentrated, thus focusing on the potential for exploitive behaviour, particularly concerning labour (Kalecki, 1939). This has had some influence in the industrial economics field where monopoly has been studied.

While mainstream economics has been a formidable tool with much empirical relevance, it has been subject to criticism (Eggertson, 1990) because it does not explain the rationale for different forms of economic organization and the effects of social rules on behaviour and outcomes. Institutional economics, although essentially adhering to the basic tenets of neoclassical theory, introduces information, time and transaction costs and property rights as constraints on the attainment of business objectives. It also emphasizes the need for empirical testing of hypotheses concerning such constraints. To this extent, the stance of institutional economics is fairly close to that adopted by much of industrial economics.

The institutional perspective on business activity is a relatively new one and for this reason there is, as yet, no clear agreement on its principles. Because it is still at the exploratory stage, there are differences of opinion as to its postulates. For example, some institutional economists reject the profit maximizing and rationality principles, replacing them with Simon's (1957) satisficing concept. The danger is that abandonment of too many principles leaves industrial economics with no theory so that any investigation is merely descriptive. Notwithstanding its infancy and the problems of the choice of theoretical foundation to adopt, institutional economics offers insights into the impact of variables hitherto ignored and it also underlines the need for economic analysis to consider market dynamics and the uncertainties encountered in the business environment. In this respect, it is consistent with game theory analysis. It also links conventional analysis with behavioural approaches which attempt to relate human activity within business organizations to economic outcomes. Furthermore, it contributes to explanations of why market failure occurs, a factor of importance when environmental issues are examined, especially regarding property rights.

Important contributions to the analysis of market imperfections in the form of information and transaction costs have been made by R. Coase (1960), O.E. Williamson (1985, 1986) and North (1990) and are reviewed by Stiglitz (1989) and O.D. Williamson (1989). These are the costs incurred in searching for and procuring information, which helps to reduce uncertainty, and in executing transactions. They are likely to be substantial in some spheres, such as where a firm assembles a product from many sources, a feature of tour operators marketing package holidays. Such imperfections provide an incentive for economic integration between firms, as in the case of the Thomson Corporation, as a means by which firms can internalize the costs which they would incur by operating as separate units. Transactions theory also facilitates the study of the relationship between principals and agents, a relevant feature in this chapter. A principal–agent relationship exists where one party's welfare depends upon the actions of another. The principal is normally affected by the actions of the agent. It is possible for this kind of relationship to occur within a firm, such as between an employer and employee or shareholders and management. Transaction costs are relevant in the principal–agent link as it may be necessary for the principal to monitor the agent's behaviour to ensure that the former's objectives are attained. Within the tourism industry there is a principal–agent relationship between tour operators and travel agents. Commission rates offered by tour operators are conventionally 10 per cent in the UK but if agents reach target numbers of holidays sold, override commissions are paid. Incentives are also provided for devoting a certain proportion of racking space for the principal's brochures and direct reservation links which cut costs.

Another important development of a specific nature, related to both transaction costs and uncertainty, has been the investigation of innovation, especially the role of research and development, for example, Scherer (1967), Stoneman (1983), Davies (1989b) and Reinganum (1989). Clearly not all research and development yields marketable products or usable processes, such as 'just-in-time' and 'total quality control' systems, and the costs of their discovery and adoption are subject to much uncertainty. There is the additional problem of the extent to which a particular firm can safeguard any discovery in order to reap the rewards of its cost of development. Attempts to model innovation and growth have been made within the neoclassical framework (for example, Dasgupta and Stiglitz, 1980; Grossman and Helpman, 1991; Aghion and Howitt, 1992, 1995).

An interesting feature of the evolutionary and institutional branches of industrial economics and the associated concepts of transaction costs, principal–agency analysis and the innovation process is that they identify market phenomena which are external to individual businesses and industries. The issue which is emerging in these areas of investigation, which is relevant to the development of industrial economics, is the extent to which the problems can be internalized to minimize their adverse impact, in terms of costs, on a firm's or industry's operation. This raises a matter of fundamental importance in economic supply analysis, namely the organizational arrangements which are necessary. In short, it poses the question as to why firms exist. Are they merely entities for transforming inputs into outputs or are they production–distribution units which, in order to operate efficiently, require organizational structures by which their objectives are attained? Moreover, is it possible that managers and employees set their own objectives which the organizational structure also facilitates? The latter notion acknowledges the force of the longer-established behavioural analysis of firms and individuals within them (considered on pp. 101–2). Such issues cannot be pursued here but a not implausible inference to draw is that although the theoretical development of evolutionary and institutional economics, transaction costs, principal-agency and innovation concepts, is in its infancy, a degree of consensus, if not convergence in economic thinking, is occurring (Dietrich, 1994). This might suggest that a more dynamic and unified theory of organizations will emerge.

Research in industrial economics does indeed demonstrate a concern with establishing a firmly rooted theoretical base (Davies *et al.*, 1989; Schmalensee and Willig, 1989; Basu, 1993; S. Martin, 1993). This is the justification for using such terms as New Industrial Economics and New Industrial Organization. It is not entirely coincidental that these movements were reinforced by the emergence of game theory as a means of analysing business strategies. Not only did the application of game theory allow for the incorporation of uncertainty and asymmetric information, but also it made it possible to construct more dynamic models, thus increasing their explanatory power. Game theory has reawakened interest in the long-established duopoly form of oligopoly models such as those constructed by Bertrand, Cournot and Stackelberg (S. Martin, 1993), included in mainstream microeconomic texts, so that the development of oligopoly theory is now a dominant element in industrial economics. This analysis has generated research into a number of issues, such as entry/exit conditions, price wars, predatory pricing, the role of advertising, co-operative contractual arrangements between the different elements of supply and demand and collusion. The re-emphasis on theory, however, has been criticized for failing to provide empirically testable models of industrial behaviour. Nevertheless, there has been renewed interest in specific case studies, relative to the earlier bias towards more generalized econometric analysis. Moreover, greater attention is being paid to the scope and impact of intervention, for example, anti-monopoly legislation, market regulation, taxation and subsidies. The role of policy and its impact on welfare have been examined within public choice theory (for example, Buchanan, 1968).

Despite the apparent conflict between the desire by theorists to develop general models of supply and the insistence by empiricists on establishing how firms actually behave, industrial economics, and with it the comprehension of what determines the pattern of supply, has made great advances. There is a better understanding of the interrelationships of suppliers within particular industries, such as between intermediate and consumer goods producers and between principals and agents. Models now provide better reflections and explanations of reality. However, most progress has been made in developing market behaviour models out of their duopoly origins, again incorporating the application of game theory and accommodating such notions as product differentiation, market segmentation, price discrimination and the reactions by firms to their rivals' price and non-price competitive strategies. Before proceeding to examine the ways in which the different approaches can shed light on behaviour in tourism markets, it is useful to consider briefly the behavioural theories of the firm which have been developed outside mainstream economic analysis.

BEHAVIOURAL MODELS OF THE FIRM

Concurrent with the development of new industrial economics but hardly acknowledged by it was the growth of behavioural models of the firm, studied within the sub-discipline of economic psychology. Behaviourally oriented theories do not rely as heavily on two major assumptions of conventional analysis, first, that business decisions are identical to those of individuals within firms, and second, that theoretically derived optimizing positions are representative of actual behaviour. Economic psychology perspectives on businesses question the axiom of rationality, perceiving it as a hypothesis which should be subject to empirical testing. The main thrust of such studies of the firm is that analysis of business behaviour produces observations which generate, through replication, broad generalizations which aid the construction of theories, through an inductive process.

Business behavioural researchers argue that decision-makers may not apply clear choice models when they lack knowledge of possible alternative courses of action and are not certain of their outcome. Preferences may be inconsistent and they may not possess firm guidelines or rules by which to make decisions. In this sense, the assumptions of conventional economic theory are challenged, as is apparent in the observations of pioneers in business behavioural theories such as Simon (1955), March and Simon (1958), March (1962), Cyert and March (1963), Simon (1979), and Cyert and Simon (1983). It is hypothesized that business people lack the information and time to optimize their activities. They are, therefore, aware of only a small number of options from which to choose. Thus, a predominant issue which behavioural researchers have investigated is the institutional structure and process of decision-making (see for example reviews by Slovic *et al.*, 1977; Ungson *et al.*, 1981; Kahneman *et al.*, 1982). Of interest has been the ways in which problems are identified, the process of learning by experience, perceptions of probable outcomes, attitudes to risk, the allocation of attention to multiple problems and activities, and organizational adaptation.

Although economic psychology, as an empirically based approach to industrial organization, concentrates on behavioural issues, it is not in direct conflict with neoclassical optimizing theory but has modified the theory by relaxing the more abstract assumptions. Its principal contributions, however, have been to widen the scope of analysis by including the human behavioural element and to strengthen the links between theoretical and empirical studies. To an extent, therefore, industrial economics and the economic psychology of the firm have grown out of traditional theory and run in parallel with each other. Recent developments in these two areas of economic analysis, with the greater emphasis on the testing of theoretical models in an empirical context, have increased their potential for explaining the structure and operation of tourism supply. Many of the issues examined by the theoretical approaches are manifested within tourism markets. Furthermore, some aspects of the structure and operation of tourism markets, especially the persistence in some markets of overcapacity and disequilibrium, question the explanations, predictions and relevance of some industrial economic and economic psychology models. Thus, it would be informative to examine the rich empirical evidence which a study of tourism supply could yield.

Hitherto, as shown in Chapter 4, with the exception of the transport sector and some examination of accommodation, tourism markets have not been analysed by mainstream industrial economics, despite the major contribution which tourism makes to the economies of many countries. Accordingly, the remainder of this chapter is devoted to an examination of the contribution which economic thinking on industrial issues can make to understanding the behaviour of tourism firms and their market conditions using the limited empirical evidence which is available.

STRUCTURE, CONDUCT AND PERFORMANCE OF TOURISM INTERMEDIARIES

The SCP paradigm is a feasible means of analysing complex markets comprising firms of differing sizes, in which varying degrees of concentration and/or integration occur and market power can be exercised (Brozen, 1971; Schmalensee, 1972; Demsetz, 1974; Cowling and Waterson, 1976; Peltzman, 1977; Spence, 1977; R. Clarke and Davies, 1982; Dixit, 1982; Lieberman and Montgomery, 1988; Tirole, 1988). The approach has the advantage of providing a clear framework which avoids unstructured descriptions but instead permits the examination of markets in terms of the analytical categories of market structure, firms' conduct and performance. The ability of the SCP approach to encompass many of the elements of tourism supply is indicated by Figure 5.1, based on Scherer (1970) and derived from Mason (1957).

In addition to showing the main characteristics of structure, conduct and performance, Figure 5.1 makes reference to public policy and its impact on firms' behaviour. The principal modification to the conventional SCP diagram, for the purpose of applying the analytical framework to tourism, is the omission of elements concerning the supply of tangible products, only peripherally relevant to a service sector, and decreased emphasis on the welfare considerations of performance which are not the focus of this chapter.

Market structure variables which are commonly used in industrial economics analysis were introduced in Chapter 4. Two which are key measures of competitive conditions are the number and size of firms and indices of

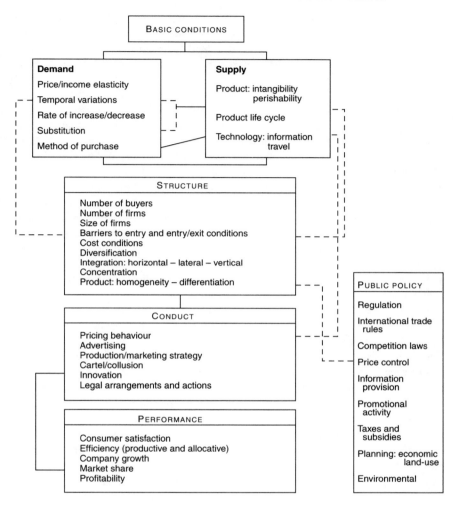

Figure 5.1 The structure–conduct–performance paradigm applied to a service industry

concentration. Additional variables which can be considered, included in Figure 5.1, are the number of buyers, entry and exit conditions, cost conditions product differentiation and diversification and inter-firm integration. Firms' conduct relates to pricing behaviour, advertising, marketing, research and development and innovation, sometimes in the context of tacit collusion or more formal cartels. Although innovation is partly seen as the result of exogenous technological change, it is also driven by competitive conditions in markets, for instance, in tourism, the pressure on suppliers to install central reservation systems or electronic means of advertising products or payments transmission in order to reduce costs. To protect an innovation or to control the sale of products, companies resort to such strategies as licences and contracts organized through the legal system, and such arrangements often reinforce entry barriers and, thus, the structural characteristics of markets.

Performance can be considered in terms of consumer satisfaction, efficiency of operations, the growth rates of firms and industries, firms' market shares and profitability. Within tourism supply, short-term performance measures have sometimes been paramount, such as in the UK package holiday sector where the concern has been for the growth of sales and market shares, often at the expense of efficiency and profitability. Performance is affected by public policy, especially changes in regulation, international trading arrangements and competition laws which have played an important role in tourism. Price control has exerted a strong influence in the transport sector, such as in the setting of international air fares, which governments have implicitly, if not overtly, supported. Furthermore, the promotion of tourism by public sector bodies and the provision of subsidies and/or tax incentives have had a marked impact on production, for example in relation to the supply of tourist accommodation.

The interrelationships between structure and the conduct and performance of firms within particular markets, as well as the basic conditions of demand and supply and public policy, are indicated by the connecting lines in Figure 5.1. The solid lines show the causal links as originally posited in the economics literature, running from structure to conduct and performance. However, the more recent formulation of the model does not deny the possible influence of conduct and performance on structure, thereby allowing its endogenous determination. In this respect conduct, where it embodies the actions of managers and employers in firms, reflects the importance which institutional economics attaches to the influence of human behaviour on outcomes. Therefore, later considerations of SCP analysis suggest, as indicated by the broken lines in Figure 5.1, that not only performance and conduct, but also public policy, affect market structure. For example, deregulation has influenced the conduct and performance of airlines and bus and coach companies and, with the passage of time, their market structure as trends towards concentration emerged.

THE TOUR OPERATIONS SECTOR WITHIN THE SCP FRAMEWORK

It is clear that a wide range of tourism supply issues can be examined in the context of the SCP framework presented in Figure 5.1. The insights which the SCP model can provide will be illustrated using the UK tour operator sector because of the sector's central position in tourism supply and the availability of evidence on its structure and performance. An overview of structure, conduct and performance characteristics, based on available data for UK tour operations, is provided in Table 5.1.

Table 5.1 shows that the structure of UK tour operations is characterized by a large number of firms, of greatly varying size, and a large number of buyers. Although barriers to entry and exit are generally low, the concentration ratio is high in terms of market share. There is little empirical evidence concerning the extent of economies and diseconomies of scale and economies of scope, capital indivisibilities, fixed capacity and fixed costs. The larger operators in the top tier, which have invested in charter airlines and

Table 5.1 An examination of the UK and US tour operations sector within the SCP framework

Structural elements	Evidence
• Number of buyers	Very many, e.g. 17 million plus in UK (1994)
• Number of firms	Very many, e.g.　1,500 + in USA 1,000 + in UK
• Size of firms	Wide range – indicated by 　• capacity (number of holidays) 　• market share (see below)
• Barriers to entry	Generally low sunk costs except for large and integrated entities; ease of entry/departure indicated by 　• birth rate:　USA 100% in 10 years to 1992 　　　　　　　UK 100% in 8 years to 1993/94 　• death rate:　USA and Europe 70 % in 15 years to 1993/94
• Cost conditions and structure	• Relatively high fixed costs of contractual arrangements but offset by 'get out' clauses • Potential for economies of scale and scope
• Diversification	Generally low except for largest operators as part of conglomerate
• Integration	High: largest own aircraft, hotels, travel agencies, facilities, e.g. Thomson, First Choice, Airtours
• Concentration	High as measured by market share e.g.　•USA 10 largest control 30% + (1992) 　　　•UK　4 largest control 60% + (1994) 　　　　　10 largest control 70% + (1994)
• Product characteristics	Heterogenous: differentiation, price discrimination, segmentation strategies
Conduct	
• Pricing behaviour	• Governed by relatively high elasticity of demand • Recurrent price wars (not necessarily initiated by dominant company) aimed at 　• filling fixed capacity 　• securing market share
• Advertising	• Relatively high percentage of sales and costs • Persuasive rather than informative

Table 5.1 (contd.)

Structural elements	Evidence
Conduct (cont.)	
• Production/marketing strategy	Objectives 　• high volume sales 　• company growth 　• product differentiation 　• market segmentation
• Cartel/collusion	Virtually none
• Innovation	• Information technology concerning reservations sometimes outside control of the sector • Some incentive for product innovation • Few to benefit individual company
• Legal arrangements	• Principal–agent legislation • Some franchising, management arrangements carry legal status
Performance	
• Consumer satisfaction	Moderate 　• governed by the potential of the many components of the product to cause dissatisfaction 　• policy of securing consumer brand loyalty
• Efficiency	Moderate 　• gross margins often less than 10% 　• driven by fierce price competition
• Company growth	• Difficult to maintain by established companies; relatively stable among top 5–10 firms • Rapid for innovative and/or early phases
• Market share	Pursued vigorously only by top tier of around 20 operators
• Profitability	• Volatile coinciding with boom in economic cycle • On average less than 5%
*Public policy**	
• Regulation	• Licences specifying number of holidays/air travel required • Bonds to safeguard consumers' interests • Deregulation of transport has affected its cost and availability
• Competition laws	Attracted little attention to date, e.g. Thomson merger with Horizon in UK not deemed to be against the public interest

Note: * Only regulation and competition laws given in Figure 5.1 are considered here

some travel agents, appear to enjoy economies of scale and scope, although such investment has entailed the problems of indivisibility and high fixed costs with the concomitant fixed capacity which needs to be filled. Average load factors on aircraft and occupancy rates in hotels are measures which indicate the extent to which such problems are overcome. However, the majority of tour operators are relatively small and these structural characteristics tend to be less important. Operators which organize the more specialized forms of tourism where the market is not extensive do not require high investment in fixed capacity and, indeed, would be likely to suffer diseconomies of scale if the more personal bespoke service were undermined.

Tour operators' conduct involves recurrent price wars and high volume sales by the largest operators in the absence of collusion between them. Firms have differentiated their products in the context of segmented markets. Performance outcomes include moderate consumer satisfaction and moderate efficiency of firms' operations, with low profit margins and some bankruptcies. The sector has been subject to regulation although anti-competitive conduct has not been deemed a significant problem to date.

In addition to providing a framework for the examination of market and firm characteristics, the SCP model predicts that structural features affect firms' conduct and performance. Owing to the important role which the structural base is hypothesized to play, two structural features which are of particular relevance to tour operations will now be examined. First, the role of entry and exit conditions for the sector will be discussed. Measures and changes in concentration will be considered subsequently.

Entry and exit conditions for tour operations

The high growth rate in the number of tour operators in the UK is indicative of low entry barriers. Capital costs are low and the identification of a niche and expenditure on promotion to bring the product to the attention of the potential tourist do not constitute major obstacles to entry into the market. It is necessary to convince purchasers of the viability of the company in terms of guaranteeing the security of payments made in advance. A significant cost of entry is therefore the bonding requirements. Additional entry barriers may also occur. Salop (1979a, 1979b) distinguishes between innocent barriers and strategic barriers which incumbent firms erect. Even if perfect information is assumed, economic theory can show that there are advantages to incumbent firms. For example, incumbent firms may have erected pre-entry barriers by having established licensing and franchise schemes, created a differentiated branded product, secured consumer and supplier loyalty and contracted with hoteliers and facilities in the most desirable locations. These kinds of asymmetries between the positions of entrants and incumbents were originally seen by SCP advocates, such as Bain (1956), as entry barriers but more recently they have been interpreted as first mover advantages (Lieberman and Montgomery, 1988). Post-entry barriers such as economies of scale and scope and lower input prices can also be identified and may confer absolute cost advantages.

Strategic entry barriers are a conscious effort by established firms to deter new firms. Typical ploys are to adopt limit pricing, perhaps of a predatory nature, to differentiate products, increase advertising expenditure in a counteractive manner, seek improvements in efficiency or capacity to lower unit costs of production. Firms have also reinforced the strategic core of the business. The strategic core is the principal activity of a business enabling it to achieve its aims and objectives, taking account of the prevailing opportunities and threats and its strengths and weaknesses. In economic terms the core activity facilitates the attainment of increased economies of scale, or extended product differentiation and multi-market targeting to exploit economies of scope. In some cases, incumbents maintain spare capacity so that they can increase output and decrease their price rapidly if a new firm enters the market, making its position unprofitable and forcing it out of the market. The maintenance of spare capacity is, therefore, a strategy which involves a credible threat to potential entrants, backed by the 'punishment strategy' of a price war. Vertical integration involving investment in sunk costs, i.e. those which are large and not recoverable in the short run without involving the business in considerable expense, such as the purchase of aircraft or a chain of hotels, can also constitute a 'pre-commitment' which acts as a deterrent to entry and exit, especially where substantial cost savings are achieved. Entry deterrence strategies and their likelihood of being successful are examined within a game theory context later in this chapter.

Tour operators have not vigorously pursued such strategies to deter entry *per se*. They have tended to be more concerned with increasing market share, accepting that they have insufficient power to exclude new entrants. In this respect, economic theory would suggest that sunk costs are low and similar for newcomers and incumbents alike. Baumol (1982) argued that low or recoverable sunk costs make markets contestable, concluding that if prices are competitive and not far above marginal cost, then monopoly and oligopoly structures are largely benign. Such a view, prima facie, appears to apply to the UK tour operations sector up to the mid-1990s, as indicated in Chapter 4. It is clearly dominated, in terms of market share, by a few firms which do not always earn supernormal profits.

However, an assumption of low sunk costs in all sectors of a market is an oversimplification. It is more likely that the level of sunk costs varies, depending on the market segment into which entry is sought and also on the strategic group (defined on p.109) into which a particular firm falls. Entry/exit conditions do not necessarily relate to an entire industry or market but may be specific to sectors or segments within them. An example serves to illustrate not only entry/exit conditions but also their interrelationship with sunk costs and economies of scale and scope. Suppose a tour operator in a specialist segment with a limited market, say activity holidays, contemplates expanding into the mass package market. Exploitation of economies of scale and scope and so the reduction of unit costs may act as an entry condition which must be met. To achieve such economies the operator may be compelled to operate its own airline and hotels, thus incurring substantial sunk costs which act as a deterrent to entry.

The strategic group argument is that in a comparatively diversified market in terms of firm size and importance and products, there are marked similarities in the characteristics of a number of companies and their products, which suggest common interests and thus that they can be classified into identifiable and distinct groups. In tour operations it has been posited (N. Evans and Stabler, 1995) that there are three tiers of companies distinguished by size, capitalization, market share and range of products offered. It is useful to re-examine this feature in order to illustrate the concept of the strategic group. The first tier, consisting of around ten operators, is more integrated and heavily capitalized and serves several market segments. It enjoys economies of scale and scope and wield considerable market power. The second tier, of around twenty to thirty firms, is more specialized but still has a significant share of the market. The remainder, in the third tier, are numerous, small and are often unincorporated, with low sunk costs. The apparent absence of barriers and deterrent action to entry bear out the contestability thesis, even in the top tier of the market; of the top five operators trading in the UK in 1994, only two were in existence in the early 1970s. However, it does not account for the presence, in the first tier, of a high degree of concentration, another aspect of market structure which has been cited as indicative of conduct and performance and a significant characteristic of tour operations.

Concentration

In the SCP framework, much effort has been devoted to determining the best method of ascertaining the extent of concentration as a measure of the competitiveness of markets. High concentration would suggest an oligopolistic structure while a low incidence would indicate a highly competitive market. However, there is a problem in using relatively crude measures, such as the number of firms. It is conceivable for a few firms to be engaged in fierce competition with each other while many firms, because they are spatially separated or serving different segments of a market, might be relatively uncompetitive, representing the Chamberlinian notion of monopolistic competition. It is, therefore, necessary to take account of the relative size of firms, as well as price, cost, profit levels and market shares, to establish how competitive the market is. Even factors like the extent of linkages between firms have a bearing on the degree of concentration.

Tour operators, particularly in the package sector, normally require a licence in order to gain access to means of travel so that it is possible to estimate reasonably accurately the number of firms. On the other hand, size may be defined in several ways, as shown in Chapter 4. Simple concentration measures, relying on size in terms of market share by number of holidays authorized by licence or actually sold, and number of firms, can easily be applied to tour operators. In the inclusive package holiday supply sector which dominates the UK international tourism market, a key measure used to indicate concentration is market share, derived from the number of holidays sold. Information about such variables as capitalization, the number employed and turnover is not readily available for small firms. Taking the UK and US tour operations sector as examples, in the USA in 1993/94 the top forty companies controlled 30 per cent of all holidays sold. In the case of UK tour operations the degree of concentration has been more dramatic, increasing markedly over time as the share of package holidays sold by the top five companies has risen to 62.5 per cent of the total of 15.5 million sales compared with 49.5 per cent in 1983.

Another example of concentration in tourism is the airline sector. Data for three variables can be used to measure market share as an indicator of concentration: sales revenues, profits and passenger numbers or passenger kilometres. However, because of the extent of both regulation and state support for national flag-carriers, the use of statistics for revenues and profits in the international market are not always feasible indicators. Indeed, in the early 1990s profits would not be an appropriate measure as, with the exception of British Airways, the largest ten airlines, in terms of passenger kilometres, all made losses. As suitable data for tour operations are not available, data for the largely unregulated US domestic air travel sector have been used to illustrate concentration. A concentration curve for this sector is given in Figure 5.2 which depicts the size of firms by using the number of passengers carried as a proportion of total numbers travelling with the top twenty-five US airlines. Thus, for example, American Airlines carried 18 per cent of all air travellers in 1993/94; the top eight airlines accounted for over 90 per cent of traffic in the same year.

Conduct and performance in UK tour operations

Interrelationships between tour operators in the UK have tended to be antagonistic, as witnessed by periodic price wars, heavy discounting and commission rate incentives offered to travel agents retailing holidays. Low entry costs, into not only tour operations as a sector but also specific segments within it, indicate the possible recurrence of price wars, as tour operators attempt to maintain or increase market share and gain scale economies by means other than formal cartels or collusion. Large companies have pursued expansion in growth and market share, more to secure their own positions in an unstable and volatile market than to oust rivals. The largest tour operators experience legislative constraints on anti-competitive strategies and also face competition from second tier tour operators. However, although the 'second tier' operators showed the most growth in the early 1990s and some of these operators can provide the element of choice that customers demand, by the mid-1990s they were losing ground in both the

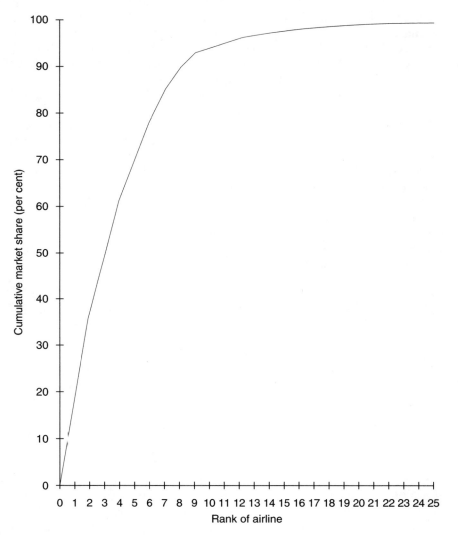

Figure 5.2 Concentration in the US airline sector as indicated by the number of passengers carried by the top twenty-five companies

mass and specialist holiday markets. This was unanticipated since they appeared to be well placed to serve the needs of both the multiple and independent travel agents (East, 1994).

Strategies undertaken by the top tier to consolidate its share of the mass market and break into the specialist segments indicate that a large number of the second tier tour operators, together with many operators in the third tier, face a high level of risk. Some companies are badly exposed, since they sell standardized mass market package products in a market that is, to a large extent, controlled by the largest vertically integrated companies which can undercut them on bed prices negotiated in bulk. Kirker (1994) argues that they are destined to fail without clear product differentiation and a clearly defined market segment. However, the 'new' and more affluent tourist is demanding more innovative, flexible and tailored products and small to medium-sized operators can meet these requirements. The chase for market share and for the 'old' mass market tourist is no longer a viable strategic option except, perhaps, for the five to ten largest tour operators and even they are likely to continue to experience cycles involving low or negative profits.

In terms of performance, there is little evidence that the leading UK tour operators have earned high rates of profit up to the mid-1990s (N. Evans and Stabler, 1995). Within the tour operator sector, profits have been relatively low in the face of variability in demand, low entry costs, the high birth–death rate of firms and a high level of price competition. What, therefore, may tour operators do to stabilize the market and ensure long-run profitability? Devices such as incentives to persuade agents to be sole distributors, restricting universal information technology systems, franchises, licensing and reciprocal agreements may be considered but some measures are difficult to enforce and others are regarded as anti-competitive by the regulatory authorities. Product differentiation, niche marketing and price discrimination are alternative strategies which may be more effective. Tour operators may attempt to deter entry into the sector, as indicated earlier in the chapter, by convincing potential competitors that entry will be unprofitable. Entry deterrence tends to increase concentration, with all the advantages of scale economies and market power it can confer. However, tour operators are likely to be constrained by a number of countervailing forces.

Ryan (1991), using a Delphi technique to assess the views of twenty-eight leading industry practitioners, identified three main constraints on further increases in concentration: first, if competition is based on quality of product rather than price, small tour operators can still compete; second, the threat of anti-monopoly legislation may act as a deterrent to strategies aimed at curtailing competition; third, large tour operators may be insufficiently adaptable and flexible to maintain quality for clients. Notwithstanding these constraints, Ryan also found that the major tour operators can establish subsidiary companies to offer specialist products and these companies can reap the benefits of economies of scale in administration and transport. In such a situation, even though the market may be shifting towards niche marketing, smaller specialist companies and greater customer awareness, there is little to inhibit growing concentration ratios.

What other strategies may large tour operators adopt to expand or safeguard their businesses? They can look for overseas expansion to fuel growth. Thomson, for instance, held talks with Scandinavian Airlines System (SAS) regarding the acquisition of Sweden's largest tour operator, SAS Leisure

(*Outbound Travel Industry Digest*, 1994), before the company was sold to Airtours in June 1994. Airtours increased its capital by sales of shares to Carnival Corporation, the largest US cruise line (Blackwell, 1996). Ultimate Holidays raised £3 million in venture capital to create a pan-European company (*Travel Trade Gazette*, 1994).

The characteristics of price competitive conduct and relatively low profits performance imply that UK tour operators have been operating in consumers' interests. This outcome has persisted over time and appears to be at odds with the implications of the SCP model that increasing structural concentration would result in higher prices and increased profits. One possible explanation is that supply characteristics should be considered within an international context. Thus, although tour operations may be construed as oligopolistic in domestic markets, such as those of the UK, USA, Germany and the Netherlands (Fitch, 1987; Sheldon, 1986, 1994), internationally no single firm or group of companies is dominant (Monopolies and Mergers Commission, 1989). Economic integration between firms will be considered further in Chapter 6, in the international context of tourism supply. However, it is clear that firms' conduct, such as their strategies towards inter-firm integration and expansion, can affect the market structure. The question which is posed in the immediate context, given the absence of uni-directional cause and effect in the SCP model, is the adequacy of the approach as a framework for analysing tourism supply.

An evaluation of the SCP paradigm in the context of tourism markets

The SCP framework is well suited to examining the characteristics of market structure and firms' conduct and performance, and allows for considerable flexibility in terms of the wide variety of variables which it can accommodate. It also provides a useful framework in which to describe industries and markets as a basis for analysis. Major criticisms of the paradigm are that market structure is assumed and that it is set too firmly within a neoclassical static equilibrium framework, resulting in limited ability to accommodate market processes. It is certainly true that having grown out of conventional market analysis, the approach has been used to ascertain the conditions for highly competitive, oligopolistic and monopoly markets to exist and its application has sometimes been restricted to focus on these objectives. The SCP paradigm in its original form does not claim to show the process of change in markets so that it is not well placed to explain, in particular, how certain market structures come about, the impact of entry barriers, the size and number of firms and the effects of firms' growth rates on market structures. To this extent it reflects the origins from which it has developed. The problem of identifying causal linkages largely flows from the model's static equilibrium analysis and assumption of an exogenously determined market structure which determines firms' conduct and performance. This has inhibited consideration of the impact of conduct and performance on structure, so determining it endogenously.

More recent investigations which attempt to take account of the dynamic nature and uncertainty of markets have emphasized the cost structures of individual firms, including those associated with acquiring information and conducting transactions. In particular, the impact on the firm of the nature of the fixed assets required, the timing of the purchase of variable inputs, the linkages between inputs and outputs, scale and scope and the range of outputs have been investigated. Such studies have departed from the traditional SCP framework, which tends to concentrate on an entire industry and to focus on product as opposed to factor markets.

The SCP paradigm should not be viewed as a fully adequate analytical framework but as a starting point for examining key economic issues. Its strength is that it is an appropriate method for providing a market perspective in contrast to one based on industry alone. It has the merit of giving a holistic view and identifies the wide range of variables which require examination. It highlights the importance of certain characteristics, for example entry conditions, which have a bearing on the number and size of firms and thus the likely contestability of the market, firms' conduct in terms of pricing behaviour and strategies and profitability. None the less if, as posited, tourism markets are not only complex but also in disequilibrium, then more recent developments in the economic analysis of industry which accommodate their dynamic nature should also be employed. Indeed, economic explanations of the tourism industry and the strategies of businesses within it need not be confined to a specific framework of analysis, such as the SCP paradigm. It is likely that the range of schools of thought, discussed at the beginning of the chapter, can offer valuable insights into the operation of tourism markets. The dynamic processes according to which firms behave, changes in market structures and a number of the insights provided by diverse schools of thought can be taken into account by game theory, whose relevance to tourism will now be discussed.

GAME THEORY AND TOURISM

Conventional analysis of supply and market structures has proved inadequate for explaining strategic interrelationships between firms, particularly within the context of oligopoly which is an increasingly prevalent form of competition. In contrast, game theory lends itself to many circumstances which arise in oligopolistic markets, particularly where uncertainty prevails, and can be used to examine many of the strategies which are employed by tourism firms. There are many situations in which decisions are interdependent and firms may gain by engaging in co-operative strategies, such as those which previously characterized the airline sector, or they may decide that competitive strategies are required, as in travel agent operations where a firm may gain an advantage by continually changing strategies. What is not in dispute, however, is that many firms in tourism supply sectors take account of the behaviour of other firms in the market when deciding upon their own strategies. The crucial question is when it pays to collude as opposed to competing.

Game theory can be used both to explain behaviour and to predict the outcome of strategies regarding price-setting, product choice and differentiation, advertising, capital investment, mergers and takeovers, and entry deterrence. In observing firms' behaviour and interaction, it is also possible to understand how markets operate and evolve, an important attribute in analysing some tourism sectors which are newly emergent and unstable. It is of value not only to distinguish co-operative (collusive) and non-cooperative (competitive) games but also to vary the initial assumptions concerning participants' behaviour, objectives and knowledge of their rivals' reactions. Normally, in simple explanatory simulations, rational behaviour by competitors, profit maximization as an objective and, initially, perfect information and symmetry are assumed, so that firms have full knowledge of their rivals' cost structures, production levels, prices and so on, as well as the market conditions under which they operate. Such restrictive assumptions can be relaxed in more complex situations and game analyses.

In addition to co-operative and non-cooperative games, it is possible to consider either 'one-shot' or repeated games and, furthermore, distinguish between those played simultaneously and others which are sequential. In some circumstances firms, mindful of the illegality of collusion in some countries, may initially conduct non-cooperative strategies but in the light of the experience of rivals' reactions establish stable patterns of behaviour which may result in tacit collusion. For example, maintaining high prices in the knowledge that rivals will do the same would confer advantages on all firms. One-shot games reflect situations where once and for all decisions need to be made, for instance whether to invest in capital to produce a new product or to deter entry. Conversely, repeated games are indicative of circumstances where continual jockeying to gain short-term advantages would be appropriate. For example, pricing and product differentiation strategies may be changed frequently so that rival firms periodically engage in non-cooperative games. There are a number of instances in tourism where this occurs, such as airlines, ferry and tour operators, engaging in price wars to increase market share, while product differentiation is attempted in order to extend a firm's total market or serve a new segment.

A distinction is made between simultaneous and sequential games because they represent different market situations calling for changes in strategies. In some circumstances, for example the Cournot duopoly model, firms decide on their levels of output at the same time, whereas in Stackelberg analysis one firm makes a decision independently, to which the rival reacts. This is an important factor in situations where the first mover may gain an advantage. In the context of tourism, simultaneous decisions are often made in the light of short run changes in demand levels and patterns in the airline, ferry and hospitality sectors, especially where there is excess capacity. Sequential or reactive strategies occur in these sectors and in the tour operator market where lead times are rather longer. For example, in the summer sun package holiday sector, the offer of discounts for early booking by one operator is almost certain to lead to retaliation by rivals if it appears that their market share will be lost.

It has already been shown in Chapter 4 that some sub-sectors of tourism supply are oligopolistic and therefore strategies concerning price setting, level of supply, product differentiation and branding, market segmentation, advertising, innovation and entry deterrence are common. In addition, it is possible to consider circumstances where a key firm can exercise a dominant strategy, perhaps because of a particular attribute the firm has such as a brand name commanding a degree of loyalty, which might give it an optimal strategy independent of its rivals. In order to illustrate the application of game theory to tourism, the cases of advertising, the pricing decision and entry deterrence are considered, using examples of strategic decisions which those in the tour operator sector of tourism supply might face. The first shows a non-cooperative, simultaneous game in which the impact of advertising on market size and share is demonstrated; the second, on pricing, is an example of a non-cooperative repeated game, while the third is a one-shot case which is non-cooperative. The second and third, however, show the possible outcomes of co-operative behaviour.

The simplest condition to examine is where firms have a dominant strategy, i.e. an optimal position can be achieved by an enterprise irrespective of what its rivals choose to do. This basic notion in game theory is illustrated by the case of a tour operator deciding whether to advertise. The example presupposes that advertising is both competitive by each tour operator to capture a larger market share as well as informative to extend the market and increase the pay-off for both firms. Figure 5.3 shows a typical 2×2 matrix in a two person (firm) game, which represents the duopoly form of oligopoly, in which the pay-offs to tour operator X with respect to advertising or not advertising (rows) are given in bold while those for tour operator Y (columns) are in italics in parentheses.

Tour operator X will advertise as this is the best strategy (best net pay-off) which can be adopted irrespective of what Y does. If Y does not advertise the pay-off for X is 25, whereas if Y does advertise X will achieve a pay-off of 20. Likewise for Y, the pay-off is respectively 10 if both advertise and 15 if X does not advertise. It can be seen, therefore, that both will advertise where the total market pay-off is 30 (20X, 10Y) in the top left cell. This position is a stable one.

		Tour operator Y	
		Advertise	Not advertise
Tour operator X	Advertise	**20** *(10)*	**25** *(0)*
	Not advertise	**10** *(15)*	**15** *(5)*

Figure 5.3 Advertising: dominant strategy case

If X does not have a dominant strategy then the optimal decision depends on what Y does. For example, in the advertising case given in Figure 5.3, if the pay-off to X is 40 if neither advertises (bottom right cell) then X's strategy is determined by what Y does. If Y advertises then so must X but if Y does not advertise, neither should X because the pay-off is much greater in adopting this strategy. Therefore, X must guess what Y will do. As Y has the same dominant strategy as before it will obviously advertise because whatever X does this is its best action. If X correctly guesses Y's action, which it should given the assumptions of the model, then a stable equilibrium can still be attained. Where a dominant strategy does not exist, perhaps where advertising would extend the market rather than increase market share for individual tour operators, the question as to whether a stable equilibrium can be reached is raised. This is especially so if the assumptions of rational behaviour and correct interpretation of rivals' strategies and actions are relaxed. Game theory can show that there might be several equilibria or that there are none. Thus, it is possible to explain why instability may occur in certain industries in which particular circumstances prevail.

The example of the dilemma facing firms in deciding on price in each trading period can be employed to illustrate the point. At about the turn of the year tour operators launch brochures for the mass summer sun market for the coming season. Given the extent of the market and the fixed capacity of airline seat and accommodation capacity, which might become excess capacity in the shoulder months, individual tour operators recognize that consumers not only are price-sensitive but also may withhold purchase in the hope of late-availability bargains. Tour operators may consider, therefore, offering holidays at low prices to induce consumers to book early and so fill capacity. However, given the tight margins within which tour operators trade, such a strategy may result in low profits or even losses. The preferred strategy would be for all to charge a high price but there is no certainty that rivals would adhere to an implicit understanding either at the beginning of or throughout the season.

Repeated games can represent this real-life situation and feasibly predict what the outcome is likely to be, the main objective of simulating the market in this way being to identify the most robust strategy. Consider Figure 5.4 which includes negative as well as positive pay-offs and is akin to the 'prisoner's

117

Tour operator Y

		High price	Low price
Tour operator X	High price	**10** *(10)*	**−10** *(20)*
	Low price	**20** *(−10)*	**2** *(2)*

Figure 5.4 The pricing issue

dilemma', a case where each participant has a dominant strategy which, if pursued, produces a joint profit which is less than that obtained if both collude.

As in Figure 5.3, X's pay-offs are again shown by the first figures in bold (rows) and Y's are indicated in italics in parentheses (columns). If both charge a high price each gains a pay-off of 10 (20 in total) whereas if both charge a low price each obtains 2, giving a total of 4 (the top left and bottom right cells respectively). The matrix also shows that if X charges a high price and Y a low one X suffers minus 10 while Y gains 20 (a net total of 10). Conversely if X charges a low price and Y a high one the positions are reversed. Obviously the preferred strategy for both is to charge a high price but the issue is whether one or the other will 'break ranks' and charge a low price in the hopes of a greater gain. As long as implicit co-operation occurs both will charge a high price. If one then charges a low price the other will follow until such time as one goes back to a high price. Provided each acts rationally, can ascertain the other's strategy and believes any threats are real, a consistent pattern is likely to emerge so that tacit co-operation occurs. However, this outcome seems to be at odds with what is being observed in the UK tour operator sector (N. Evans and Stabler, 1995; Taylor, 1997) where price wars continue to occur, i.e. non-cooperative behaviour persists. The likely reason is that other factors, such as variations in demand and costs, low entry barriers, excess capacity and the presence of many firms preclude the establishment of co-operative strategies. Uncertainties introduced by such phenomena make it difficult to create the required stability.

This kind of analysis of price takes the theory of oligopoly forward from the static hypothesis of the kinked demand curve depicted in Figure 4.6 (see p.69) in the previous chapter. However, game theory can also accommodate other market conditions and situations, such as first-mover advantages, threats, including those referred to above regarding entry and the consequent change in numbers of firms, which are factors which if stabilized might, in turn, stabilize markets and therefore prices. It is not possible to examine these cases in detail but some further exposition of the implications of relaxing the assumption of simultaneous action by firms is of value.

The Cournot model of duopoly considers the case where firms decide how much output to produce simultaneously while Bertrand's approach examines a context in which firms engage in simultaneous price-setting. In contrast, the Stackelberg model posits that one firm sets output before the other. The Stackelberg approach facilitates the analysis of first-mover advantage, for example in research and development and investment, entry deterrence and marketing actions, such as advertising. Sequential games, as they are called, make game theory dynamic as it is possible to trace the processes which are initiated once a strategy is implemented. Examining product choice as an example, if two specialist tour operators, each unaware of the other's intentions, decide to introduce one of two new types of holiday, either an activity or cultural holiday with a limited market, then it is likely that both will lose money if they market the same product simultaneously. However, if firm X is able to market the activity holiday first, then Y will introduce the cultural holiday and so both will gain. This is essentially what the Stackelberg model posits, where the size of output can be determined by the first-mover, leaving the remainder of the market to the rival. Firms can ensure that they can obtain an advantage as first movers by preemptive strategic moves, i.e. by influencing rivals' choices. This can be done only if the firm has a reputation for doing what it states it will do; lack of commitment to a strategy suggests to rivals that subsequent announcements or threats as to a particular course of action are empty. Thus, for example, the threat by Thomson to withdraw the marketing of its holidays through travel agents who did not adopt its TOPS reservations system was not seen as idle.

Entry deterrence is another key strategy because, if successful, it confers greater monopoly powers and the potential to increase profits. Therefore, the incumbent firm has to convince potential entrants that it would be unprofitable to enter. This can be demonstrated by reconsidering the pricing decision as to whether to charge a high or low price. If the potential rival decides to enter the high-priced market, the incumbent firm will gain more by maintaining a high price. However, should the incumbent firm threaten a price war which would bankrupt the new entrant, and/or expand capacity to exploit economies of scale to sustain the possible price war, the deterrence commitment is substantiated. Figure 5.5 illustrates the sequence of moves and the impact on pay-offs.

In Figure 5.5, Scenario 1, the first row indicates the pay-offs at both high and low prices to the potential new entrant, X, shown in bold, and incumbent Y, shown in italics in parentheses, should X decide to enter the industry, incurring sunk costs of £20 million, it being assumed that the market is shared equally. The second row, in italics, indicates the pay-offs to incumbent firm Y if it takes action in response to X's intended entry and successfully deters it without having to incur increased costs.

Suppose firm X predicts that Y will maintain a high price in order to gain the most it can of the current £50 million total market pay-off, then X will get £5 million (£25 million half-market share minus £20 million sunk costs) and Y will retain £25 million (top left cell of the matrix). However, should Y

Scenario 1   Incumbent Y

		Profits (£m) High price	Profits (£m) Low price
New entrant X	Enter	**5** *(25)*	**–5** *(15)*
	Not enter	**0** *(50)*	**0** *(30)*

Assume: • Entry costs of X equal £20m; market shared equally
         • High price: maintained by Y after entry of X; X also maintains high price
         • Low price: price war by Y; X also charges low price

Scenario 2   Incumbent Y

		Profits (£m) High price	Profits (£m) Low price
New entrant X	Enter	**5** *(10)*	**–5** *(15)*
	Not enter	**0** *(35)*	**0** *(30)*

Assume: • Incumbent Y invests £15m to increase capacity; sales and profits
          unchanged at high price
         • Incumbent Y increases sales and profits at low price to offset cost of
          increased capacity

Figure 5.5 Entry deterrence strategy

decide to engage in a price war, i.e. charge a low price, then assuming a total market pay-off of £30 million, X will suffer a loss of £5 million (£15 million minus £20 million sunk costs) and Y will receive £15 million (top right cell of the matrix). If Y successfully deters entry by X then, at a high price, it gains all the pay-off available of £50 million (bottom left cell). Should Y threaten a price war and lower price to signify its commitment to deterring entry, then it gains the total market pay-off of £30 million (bottom right cell).

In Figure 5.5, Scenario 2, it is assumed that Y, in order to substantiate its threat to deter entry, is obliged to undertake a £15 million investment pro-gramme to increase its capacity. This may allow it to exploit economies of scale so reducing unit costs, thereby enabling it to sustain a price war should it be necessary to charge a low price. If X nevertheless decides to enter the industry and Y elects to charge a high price but is unable to extend its market or increase profits, the result is shown in the top left cell where X is in the same position as before but Y obtains a £10 million pay-off (£25 million minus the £15 million investment cost). If Y predicts that by charg-ing a low price it will sell more and at least recoup the investment cost, the result will be as shown in the top right cell where X loses £5 million and Y retains the same pay-off as before of £15 million. Of course if Y has completely misjudged the market and fails to sell more, then a zero pay-off is suffered, i.e. the £15 million share is wiped out by the £15 million investment cost; this is not shown in Figure 5.5, Scenario 2. Clearly, if X decides not to enter the industry, Y suffers a reduced pay-off of £35 million (bottom left cell) if a high price is charged and the market is unchanged so that the £15 million investment cost reduces the total pay-off (£50 million minus £15 million). In the bottom right cell the pay-off for Y is shown as £30 million if the investment pays for itself whereas if Y has misjudged the market the net pay-off will be £15 million.

The example in Figure 5.5 serves to show that many uncertainties arise when a would-be entrant attempts to break into a market dominated by an incumbent firm. The outcome for both entrant and incumbent largely depends on the extent to which deterrent threats are credible, the cost of gaining entry or deterring it and the changes in sales and profits with changes in prices. It is clear, however, that should the entrant and incumbent collude to charge a high price, the result is more beneficial to both than a price war.

Within tourism supply, as argued earlier, entry costs are often relatively low and the opportunities for pre-emptive investment are limited so that deterrence strategies are unlikely to be successful. Nevertheless, the example given here serves to illustrate one important aspect of tourism development in an international context. Investment, supported or subsidized by the state, can give an advantage to a country. For example, airline, ferry and train operations and, to an extent, the serviced accommodation sector, are characterized by substantial economies of scale. By subsidizing these, more rapid expansion can be encouraged. This might prevent foreign firms from entering the market, thereby allowing the domestic sectors to charge higher prices and/or obtain larger sales. This notion is significant in the case of developing countries which choose tourism as the engine of economic development and is considered further in Chapter 6.

This relatively brief and simple exposition of game theory, using a few illustrative examples, indicates the direction in which economic analysis of industrial operations and market behaviour is moving. Attention has tended to concentrate on competitive interaction between firms because this reflects the current position in most tourism markets. Game theory can accommodate many other circumstances, for example collusion where firms produce a low output and sell for a high price, akin to the outcome under a monopolized market structure, or instances where greater uncertainty prevails and there are a considerable number of possible outcomes (multiple equilibria). In more complex cases, the analysis involves determining the probability of specific pay-offs which accords much more closely with dynamic real-life situations.

Game theory can also accommodate different theories concerning the process by which firms reach a decision about the strategies which they adopt, put forward by the schools of thought discussed at the beginning of this chapter. A range of modes of behaviour is feasible and the Chicago school's neoclassical tenet of optimizing behaviour in competitive markets is a specific case within this range. The Behavioural school's hypothesis that decisions are made according to established norms and rules and the Evolutionary Economics school's view that behaviour evolves in accordance with past practice can also be incorporated within the range of possible modes of decision-making. The concept of updating can be used to describe consistent decision-making based on learning about other firms' behaviour (S. Martin, 1993: 588-90) and economic psychology can help to explain the process of expectations formation. Principal–agent models shed light on behaviour by owners and managers respectively, permitting them to have different expectations and reactions. Hence, game theory is capable of analysing a variety of modes of decision-making and behaviour by firms in different sectors of tourism, in different countries and contexts.

CONCLUSIONS

This chapter has examined tourism supply using the SCP paradigm and game theory and has also endeavoured to examine in greater depth a number of the economic issues in tourism identified in Chapter 4. Recent developments in the industrial economics field were introduced in order to provide additional insights into the operation of tourism markets. Underpinning the discussion is acceptance of the need to perceive tourism markets as dynamic and frequently in disequilibrium.

The SCP paradigm can be used to describe tourism markets and to highlight key features as a necessary first step towards the analysis of market structures, behaviour and performance. The advantage of the SCP paradigm is that it gives an overview showing the interrelationship of many elements. These can be added to or adapted to reflect the circumstances in different markets, as was indicated by the application of the framework to tourism intermediaries. However, the approach is not well suited to examining the process of firms' decision-making and behaviour, for which recent developments in game theory are more appropriate.

Game theory has the ability to encompass insights from various schools of thought and to provide a dynamic analysis of tourism firms, industries and markets. Thus, it can incorporate the Austrian and Evolutionary schools' critiques of static equilibrium analysis of markets. It can also consider the role of technological change, which Schumpeter highlighted as of key importance in the growth process, by showing how firms use research and development and the associated changes in technology as a competitive strategy vis-à-vis other firms. Such strategies can alter the structure of the market over the long run. Uncertainty and asymmetries in information availability, for example between market incumbents and potential entrants, can also be accommodated. The theory acknowledges that, when making business decisions, most firms consider the likely responses of competitors. Inter-firm differences in the extent of knowledge, the degree of co-operation or non-cooperation can be taken into account and simple one-off strategies or sequential adjustment to different moves by competitors can be examined. The approach can indicate the credibility of signalled intentions and the probability that threat strategies will be successful. It has the merit of ensuring that the range of strategies available to firms and the associated range of possible outcomes are made explicit, showing how firms' conduct in terms of the strategies which they use and the structure of the market are determined simultaneously. The theory incorporates the most recent developments in industrial economics and, because of its widely encompassing framework and ability to analyse the processes of change, is clearly appropriate for examining tourism market structures, conduct and performance.

The discussions of supply in this and the previous chapter have not taken explicit account of the international context of tourism markets, despite the importance of cross-country links between firms at different stages of the supply chain. The globalization of tourism supply, involving a variety of forms of economic integration, has implications for specialization in production and trade between different countries. Such issues as the contribution of tourism to income, employment and foreign currency earnings and the relationship between tourism and economic growth are relevant. Attention is now turned to these issues, prior to investigation of the wider environmental effects of both domestic and international tourism.

6

TOURISM IN AN
INTERNATIONAL CONTEXT

INTRODUCTION

The previous two chapters examined tourism within a predominantly domestic context and this chapter extends the discussion to the international arena. Tourism has been one of the highest growth activities in the world during the 1980s and 1990s, in terms of both expenditure and foreign currency generation. The high levels of tourism expenditure have significant implications for both tourist origin and destination countries, contributing to a worsening of the balance of payments of net tourism generator countries and an improvement of that of net recipients. Hence, tourism can raise or lower a country's dependence upon other countries and can be of particular importance to developing countries whose economies, apart from tourism, are based on primary products. While over the long term both domestic and international tourism can make a significant contribution to a country's economic growth, its potential for generating income and employment within a destination may be constrained by the country's ability to supply the goods and services which tourists wish to consume. Tourists' consumption of food and beverages which are imported from their country of origin, in hotels owned and managed by fellow nationals, is a prime example of the leakages of receipts from a destination. Therefore, since the pattern of tourism supply and associated distribution of receipts have considerable effects on countries' economies and welfare, it is useful to provide some explanations of the patterns which occur on a global scale. International economics can help to provide such explanations, which are the prerequisite for the formulation of policies designed to alter the cross-country structure of production. The discussion builds on the industrial economics analysis in Chapters 4 and 5 by examining tourism supply in an international context.

This chapter is divided into two main parts. The first examines possible economic explanations for international tourism, based on trade theory, and the second considers some of the effects of international tourism. The first part of the chapter is concerned with issues which have received virtually no attention in the tourism literature, while the effects which are subsequently considered have been the subject of some empirical research. The chapter starts by discussing the traditional theories of comparative advantage of Ricardo and Heckscher and Ohlin in a context of competitive market structures. Alternative facets of international tourism based on imperfectly competitive markets are then examined, including the demand for ranges of tourism product qualities, the supply of differentiated products and differences in technology. Explanations of the increasing dominance of the tourism industry by large multinational firms are provided and the different forms of cross-country integration which occur are discussed. The second part of the chapter considers the implications of tourism for income generation, employment, export earnings instability and growth. Little empirical work has been undertaken on many of the topics covered in this chapter, precluding in-depth examination of them. The aim is, therefore, to point out some of the contributions of economic research to tourism analysis and to indicate a variety of areas where economic theories have provided insights which require further investigation. Throughout the chapter the term 'developing countries' is used to denote the relatively poor nations which are elsewhere known as less developed or underdeveloped countries or the Third World, while the wealthy countries, which are sometimes termed high income or industrialized countries or the First World, are here referred to as developed countries. As D. Harrison (1992) points out, the choice of terminology is problematic. The terms used in this book are not intended to imply that one category of countries is superior to another or that all aspects of a move from developing to developed status are desirable. These issues will not be considered in the chapter but remain an important topic for debate.

EXPLANATIONS OF TRADE AND TOURISM

Comparative advantage

The vast majority of international tourist flows are relatively short distance, involving travel between geographically proximate countries for business and holiday purposes. Nevertheless, a considerable number of tourists travel longer distances to both developed and developing countries and 'long-haul' tourism has experienced significant growth as more efficient, lower cost air transport has been provided. Developed countries' foreign currency earnings from their exports of manufactured products are supported by service industry earnings including inward tourism, whereas expenditure on overseas trips by their residents results in currency leakages abroad, often resulting in a net loss on the tourism account. Many developing countries, which have traditionally relied on earnings from exports of primary products, receive net currency inflows as the result of diversifying into tourism and others are attempting to gain additional receipts by increasing tourist flows from abroad. Tourism's image as a pot, if not of gold, at least of foreign currency, raises the question of why some countries have specialized in tourism and whether gains have resulted from the new pattern of production and trade.

One of the most well-known explanations of international trade is the Ricardian theory of comparative advantage. The theory's distinctive contribution lies in its main tenet that even if one country is more efficient in absolute terms in producing goods than another, short run gains from trade can be obtained if it specializes in the production and export of the goods which it produces relatively efficiently, i.e. in which it holds a comparative advantage. If each country were to specialize, total output would be greater as a larger quantity of the goods could be produced for given inputs. The theory, thus, predicts that the pattern of trade is determined by differences in the relative efficiencies of production in different countries and that gains can result from specialization in production.

According to Ricardian theory, given competitive conditions, each country's domestic (non-trade) price ratio is determined entirely by supply-side conditions, the relative efficiency of production stemming from technology. The post-trade price ratio is determined by both supply-side conditions and by demand, based on consumers' preferences for the traded products. Departing from the static context of the theory, differences in the rates of growth of demand for the two products can result in a movement in the commodity terms of trade (defined as the price of exports divided by the price of imports); for example, against the country producing and exporting the product for which there is a low growth of demand. The country is able to purchase fewer imports per unit of its exports and therefore income and welfare fall. Thus, specialization can possibly have disadvantageous effects over the long run. The long-run effects of specialization according to differences in comparative advantage will be considered at the end of this chapter.

The Heckscher–Ohlin (HO) theorem posits that a country's endowments of factors of production (labour, capital and land/natural resources), rather than relative efficiencies of production, determine its comparative advantage. Thus, countries such as Tanzania, which have a large supply of labour and land as well as plentiful natural resources of wildlife, mountains and beaches, would appear to have a comparative advantage in tourism. The HO theorem has been applied to the agricultural and manufacturing sectors and attention has generally focused on the endowments of labour and capital. Thus, a country which is relatively well endowed with labour is said to have a comparative advantage in producing and exporting goods which are produced labour-intensively, while a country which is capital-abundant has an advantage in producing and exporting capital-intensive goods. Samuelson (1948, 1949) took the theory a stage further by arguing, in the factor price equalization theory, that trade would have the effect of equalizing the returns to capital and labour across countries. This would occur as consumers demand products with relatively low prices stemming from low labour costs, thereby increasing the demand for labour and the wage rate. A similar process would occur for products incorporating relatively low capital costs.

The HO theory is useful in so far as it points to the role which the supply-side can play in determining the pattern of international production and trade, thereby building on the closed economy analysis of tourism supply of Chapters 4 and 5. It might, at first sight, be assumed that tourism provision is labour-intensive, so that countries which are well endowed with labour have a comparative advantage in tourism. However, J. Diamond (1974) pointed out that tourism can involve large inputs of capital as well as skilled labour, as in the case of Turkey, which is supposedly a labour-abundant country. Studies of the values of the incremental capital–output ratio (ICOR) which is the number of units of capital required to produce one more unit of output or income, in the tourism sectors of Kenya, Mauritius and Turkey, provided estimates of 2.4–3.0 for Kenya (F. Mitchell, 1970), 2.5 for Mauritius (Wanhill, 1982) and 4.0 for Turkey (J. Diamond, 1974, 1977). Comparable estimates for the agricultural and manufacturing sectors of each country were 2.7 and 4.4 for Kenya, 3.3 and 3.9 for Mauritius and 2.3 and 2.1 for Turkey. Thus, the capital intensity of tourism varies between countries and can also vary over time, at different stages of tourism growth. Since tourism is not homogeneous it is likely that tourism is relatively labour-intensive in countries with a large supply of labour and capital-intensive in countries which are capital-abundant. Estimates of the factor intensities of tourism production in different countries and time periods would be useful and could indicate tourism's potential for reducing unemployment in destination countries.

The relationship between tourism production and factor endowments is further complicated by the problem of measuring factor abundance and quality. Abundance can be measured in terms of either quantity, and hence supply alone, or value, in which case demand also enters into play as higher demand for the product results in a higher price and higher value. Its quality is more difficult to ascertain as, for instance, labour can be classified as skilled or unskilled and capital can be of differing vintages and efficiency. When the third factor, land, is considered, the situation becomes even more complicated. Land in either the form of natural resources or the built environment is usually a key component of the tourism product but it is not clear whether land is complementary to capital or labour or both. Land may be complementary to capital in some types of tourism supply. For example, some tourism products within an economy which is well endowed with labour may be capital-intensive. Game tourism for high-income groups visiting Kenya, where light aircraft are used to gain access to more remote areas, is an example. Alternatively, land may be complementary to labour, as in the case of trekking holidays in Nepal.

The issue of whether a capital-intensive form of tourism production can become more labour-intensive in an economy characterized by under- or unemployed labour or, conversely, whether tourism can become more capital-intensive in a context of low birth rates, is clearly important. In Spain, for example, traditionally labour-intensive hotels have introduced more self-service in their restaurants as wages have risen, despite the context of high national unemployment rates. Changes in the factor intensity of production and comparative advantage can be advantageous or disadvantageous to a country and require consideration in a dynamic rather than static context. The concept of dynamic comparative advantage will be discussed later in the chapter.

To summarize, Ricardian theory is useful in indicating the gains which countries can make from international tourism if they are relatively efficient in tourism production and, hence, points to the importance of increasing production efficiency. The HO theorem's emphasis on the role of countries' different resource endowments also helps to explain international trade and tourism. An obvious implication of the HO theorem is that further research should be undertaken to investigate inter-country differences in the factors of production which are used in tourism and the ways in which countries might use their resources more effectively. However, the theorem's assumption that land, labour and capital are homogeneous can be called into question and relative factor endowments can change over time. For example, consideration of the differences between the natural and built environment resources used in tourism in different locations demonstrates the significant inter-country diversity of the supply of land and natural resources.

Neither Ricardian theory nor the HO theorem pays sufficient attention to the role of demand; for example, diversity of demand may occur as preferences relating to the consumption of tourism and other goods, initially discussed in Chapter 2, differ between countries. Moreover, the assumption of competitive market conditions, while offering interesting insights into the determination of comparative advantage in tourism, precludes examination of features which characterize imperfectly competitive markets. Thus, the theories of comparative advantage concentrate on supply-side differences in relative factor efficiency and factor endowments but neglect such characteristics as increasing returns to scale, market power and control over pricing or output, product differentiation, transport costs, inter-country differences in demand, market segmentation, differences in factor 'quality' and differences in countries' access to information and technology. A number of these considerations are relevant to an explanation of the high level of international tourism flows occurring between countries with similar factor endowments and similar levels of income and wealth, as will be seen in the following section.

International tourism and imperfectly competitive markets: overlapping tastes

Many tourism markets depart from traditional assumptions of competitive markets and constant returns to scale, as was shown in Chapters 4 and 5. On the supply side, tourism markets are characterized by a multiplicity of structures and tourism products and on the demand side, consumers demand a wide range of types of holidays. The fact that tourism supply involves products of particular types and consumers demand tourism products of particular qualities implies that insights from theories of international trade in imperfectly competitive markets can shed further light on international tourism production and trade. A number of features of such markets were previously considered in Chapters 4 and 5, in the context of a closed economy. They will now be discussed within the international economic arena, starting with Linder's (1961) explanation of intra-industry trade, i.e. two-way trade in products supplied by the same industry, such as exports of Rovers by the UK to France and UK imports of Renaults.

Linder highlights consumers' similarity in tastes as a cause of trade. He provides an explanation of intra-industry trade which is complementary to theories which focus on monopolistically competitive market structures and also provides a rationale for the high proportion of total tourism flows between geographically proximate countries with similar levels of income and wealth. Linder argues that suppliers from a given country initially offer a variety of products to meet demand from the domestic market. In the case of tourism, these might be, for example, holidays in different natural or built environments, in a range of accommodation types and with a choice of different sporting, entertainment and other leisure activities. Suppliers from the country subsequently provide a range of products for export, for example, holidays specifically designed to meet the preferences of foreign tourists. According to Linder, the more similar the demand for the products supplied by different countries, the greater the likelihood of trade between them.

The explanation is based on the fact that the nature of demand by consumers is determined by the level of per capita income within a country. Residents of developed countries with relatively high per capita income demand a range of higher quality products while those in developing countries are more likely to purchase a range of lower quality goods. Consequently, there is more overlap in the range of product qualities demanded by people in countries with similar income levels and less overlap in the ranges demanded by consumers and supplied by firms in rich and poor countries. Hence, there is greater potential for trade between countries with similar levels of income than between those with dissimilar levels, as the former have more market segments in common than the latter. On the demand side, trade permits consumers in countries with overlapping market segments to take advantage of a greater variety of products and on the supply side, producers benefit from economies of scale and scope in production which facilitate sales abroad. Thus, trade has the advantages of increasing variety and lowering prices (Krugman, 1980).

Linder's theory helps to explain the high level not only of trade but also of tourism movements between countries which, in contrast to the postulates of the HO theorem, have relatively similar factor endowments. The theory predicts the quality range but not the specific tourism products a country will supply. In fact, most intra-industry trade, including that in tourism, is in products which are differentiated by such means as branding, publicity and promotion, by a unique technology or by the use of a specific local environmental resource. Product differentiation provides producers with the ability to influence the prices of their products and to obtain supernormal profits in the short run, as indicated in Chapters 4 and 5 and in the discussion below. Such advantages are sometimes reinforced by restrictions on competition from foreign firms, for example, restrictions on sales of package holidays by tour operators based in other countries.

Economies of scale and scope

Economies of scale and scope provide gains from international trade, even in the absence of inter-country differences in technology or resource endowments (Helpman and Krugman, 1993). Many firms within the tourism sector experience economies of scale, as indicated in Chapter 4. Suppliers wish to achieve decreasing unit costs not only in order to increase profitability but also as a means of deterring competitors from entering the industry. However, they may be unable to attain significant economies by tapping the demand for tourism from domestic residents alone because of the more limited extent of market demand. International demand for their products can enable them to increase their output sufficiently to attain the desired increasing returns to scale.

On the other hand, relatively small tourism firms, for example, in low income countries with a low level of demand, may be unable to compete on an international scale owing to their high initial costs and prices and may be eliminated from the market before they are able to gain sufficient economies of scale to compete effectively. In such situations there may be a case for short-run protection of the 'infant industry' until it has attained economies of scale on a par with those of larger tourism suppliers. The infant industry argument has provided a partial justification for government protection of some international airlines, although an element of national pride is a factor. However, while some have increased their efficiency, others have taken advantage of their featherbedded status or been unable to achieve the economies of scope necessary for effective competition, as in the case of Cook Islands International airline (Burns and Cleverdon, 1995).

Product differentiation

Consumers' demand for diversity of tourism products is met by product differentiation by the large number of firms in monopolistically competitive domestic markets, as in the cases of accommodation and entertainments. As the demand for tourism extends from domestic to foreign markets, some firms which supply products with characteristics similar to those sold by foreign competitors cease operations, as their rivals are better placed to take advantage of economies of scale. Others, however, are able to differentiate their products from those of their rivals, for instance, by means of branding in the international hotel sector, and so remain in the market, increasing the range of products sold. Hence, the analysis of international competition under monopolistically competitive markets points to the role of product differentiation and targeting niche segments of the market, along the lines of producers of manufactured goods in Taiwan (Rodrik, 1995). Therefore, product differentiation, like the attainment of economies of scale and scope, can be used as a tourism business strategy. It also suggests that international competition can give rise to a more imperfectly competitive domestic structure as some firms are unable to compete successfully and so fewer and by implication larger firms remain within a country.

Similar conclusions can be drawn with regard to competition between firms which are national monopolies or oligopolies but which are subject to competition in the international arena. Airlines which compete within a liberalized international environment are a case in point. The move towards increased liberalization of the international air transport market has encouraged a number of airlines to compete by differentiating their product from those of the other firms in the market. For example, some airlines provide a non-stop service to the destination at a relatively high price, as in the case of Air Mauritius where the government prohibited charter flights (Burns and Cleverdon, 1995), while others include stopovers and a longer flight time but charge a lower price for the flight. Although large airlines are concerned to exploit the economies of scale associated with large arrival and departure hubs in major cities, some smaller airlines such as Trinidad and Tobago's British West Indian Airways (BWIA) can successfully supply transportation on specific routes (Melville, 1995), thereby obtaining particular niches of the market. Competition between oligopolies also takes the form of the provision of a range of qualities of a given product, known as vertical differentiation, or of products of comparable quality but different characteristics – horizontal differentiation. Thus, airlines such as Ryanair have specialized in the provision of low price flights with limited services, while other airlines have marketed flights with high quality services at a higher price. A wide income range among consumers, in conjunction with low fixed costs but rapidly increasing variable costs of quality improvements, tends to be associated with a large number of both product qualities and firms. The opposite demand and supply conditions give rise to a small number of product qualities and firms, and trade is likely to force out of the market those firms whose products are of similar qualities at higher prices. This has occurred in the accommodation sector as some domestically owned hotels have been taken over by international hotel chains which can take advantage of economies of scale and scope.

Competition via research, development, innovation and imitation

In addition to competing on the basis of price or quality, firms can compete by engaging in research and development in order to gain a technological lead over their domestic and foreign rivals. Countries which gain such a lead are said to have a technology gap over other countries and if technological progress by the leading firms is ongoing, as will be discussed in the context of growth (pp. 148–51), the leaders are likely to retain their competitive advantage and boost their foreign and domestic sales (Posner, 1961). Examples from tourism sectors include the US and west European production of aircraft and of information technology such as computer reservation systems for travel, accommodation and destination facilities. Poon (1988) argues that Jamaica's SuperClub hotels are a further instance and that their success relies on innovation rather than imitation. The open economy development of the product cycle theory (Vernon, 1966) takes explicit account of the role of factors of production in determining technology gaps and the location of supply. According to the theory, the development and supply of new products tends to occur in capital-abundant, high-wage countries, underpinned by domestic demand for technology-intensive products and a supply of relatively high-skilled and high-wage labour which is able to develop and produce them. Increasing standardization of the product, imitation by foreign producers and rising foreign demand result in a transfer of production from the initial innovating countries to medium-wage economies and subsequently to low-wage and therefore low-cost countries.

The theory is interesting in pointing to the role of both demand-side and supply-side factors in determining trade and to the process by which comparative advantage changes over time. It can help to explain the transfer of production from high-wage to lower-wage countries and it is consistent with the initial concentration of international tourism demand and supply in relatively high-income countries and the subsequent growth of foreign holiday tourism in lower-income countries, although other contributory factors such as rising incomes are also relevant. For example, it is clear that each country has a specific set of cultural and environmental endowments which preclude complete product standardization. The fact that such endowments together with accommodation and entertainments are consumed on-site, in conjunction with demand for differentiated products, ensures that tourism production and consumption occurs in both low- and high-income countries.

The tourism resort cycle model (Butler, 1980; Cooper, 1992; Getz, 1992; I.R. Gordon and Goodall, 1992; di Benedetto and Bojanic, 1993; I.R. Gordon, 1994; Douglas, 1997; Tooman, 1997) bears some resemblance to the product cycle theory in emphasizing the process by which the demand for and supply in particular tourist resorts can change over time, exemplified by resorts on Spain's Costa del Sol. However, it differs in that it is more closely related to the product cycle theory formulated in industrial economics than to the open economy version of it and is also concerned with such issues as the role of the local environment and the externalities which are associated with the growth of tourism (see I.R. Gordon, 1994, for a detailed discussion). Debbage (1990) examined the role of oligopolistic tourism firms and their strategies with respect to market share, innovation and diversification in his resort cycle model of the Bahamas and I.R. Gordon (1994) considered the effects of monopoly in accommodation or land ownership.

International interdependence of firms

As shown in Chapters 4 and 5, the key characteristic of oligopolistic competition is interdependence. At the international as well as the domestic level, tourism firms may engage in 'Cournot competition', setting their output in the light of their assumptions about the other firms' output. Alternatively, they may engage in 'Bertrand competition', setting their prices according to their predictions of other firms' prices. The assumption that other firms will not change their output or prices is known as zero conjectural variation.

The behaviour of some international airlines may be explained by the reciprocal dumping model (Brander, 1981; Brander and Krugman, 1983) which examines the effects of Cournot competition in the international arena. If national markets are segmented each state airline can act as a monopoly, setting prices independently in each market, charging relatively high prices and supplying a small number of flights. International competition changes the market structure from monopoly to oligopoly, decreasing the degree of market concentration, and each airline independently perceives the opportunity of making further profits by selling to foreign consumers. The outcome entails the provision of increased supply at lower prices. The extent to which supply rises and prices fall depends upon each airline's conjectural assumptions about the way in which the other airline will respond to its actions. If, on the other hand, the firms were to engage in Bertrand competition on the basis of prices, the outcome would be likely to be different; flight prices would remain high if each airline perceived that cuts in the prices of its flights would be matched by its rivals. Tourism firms' choice of competitive strategy is, therefore, as important in the international as in the domestic arena.

Economic integration between firms

Tourism firms may respond to competition from domestic or foreign rivals by means of integration. There may be common ownership between firms or, alternatively, firms have contractual links with each other with various intermediate arrangements between these positions such as management contracts or franchising. Three main types of integration occur between firms. Vertical integration involves the co-ordination of production by firms supplying different types of output within a production sequence, for example, Airtours' tour operations, Going Places chain of travel agents and Airtours International Airline. Horizontal integration consists of economic linkages between firms which supply the same type of output, as in the case of international hotels chains such as America's Sheraton or the Hong Kong-based group, Hong Kong and Shanghai Hotels. Conglomerate integration involves economic linkages between firms producing different types of product, as in the case of the Lonhro Corporation's ownership of international hotels or Thomson's tourism as well as printing and publishing activities. This type of integration can fulfil the function of decreasing the risk of volatility in earnings as the returns from all the different types of activities within the conglomerate are unlikely to be positively associated; thus, profitability and shareholders' returns are stabilized. Conglomerate integration can also decrease the cost of capital to the firm and a particularly profitable activity may, at times, subsidize one which requires further investment.

Vertical integration can take the form of common ownership of firms supplying different components of tourism supply but has followed different patterns in different countries. Within the UK and Spain, for example, joint ownership of travel agents, tour operators and charter airlines has been common and airlines such as British Airways (BA) and the German Lufthansa have had equity holdings in hotels in some destination countries. German tour operators generally rely more heavily on scheduled flights than UK tour operators but, unlike most UK and Spanish operators, are characterized by some ownership links with destination country hotels. Italy's national flag carrier, Alitalia, had a majority shareholding in the tour operator Italiatour and the state railway had almost total ownership of the largest travel agency in the country. There is also some common ownership of tour operators, airlines and accommodation by French firms. Conglomerate integration occurs between firms in different sectors of the economy, such as the total or partial ownership of the German tour operators International Tourist Services (ITS), NUR Touristik and Touristik Union International (TUI) by large department stores, partially owned by banks which provide a useful source of finance (Drexl and Agel, 1987).

The advantages of vertical integration include the reduction of transactions costs (R.H. Coase, 1937; Buckley, 1987; Casson, 1987) and improved synchronization of tourists' inter-country transportation, accommodation provision and entertainments (Bote Gómez et al., 1989). Vertical integration facilitates information acquisition (Arrow, 1975), the provision of inputs at known prices (Oi and Hurter, 1965) and can reduce uncertainty about future demand (Carlton, 1979). It may also be used as a strategy to increase market power (Hymer, 1976). The advantages of integration to the firm may be offset by disadvantages, including an increase in fixed costs, foreign investment risk, reduced flexibility of operation and 'dulled incentives', whereby ownership ties with an inefficient producer dull the incentive to search for a more efficient supplier. Vertical integration, in particular, can constitute the raising of an exit barrier (Harrigan, 1985), so that a firm finds it difficult to cease to supply a product for which demand has declined. A topical example is that of ownership of tourist accommodation in a destination which is experiencing long-term problems of terrorism or other forms of violence or is in decline, being on the downward slope of a destination life cycle curve. Thus, many tour operators and airlines have been reluctant to purchase hotels, even in major tourist destinations. However, foreign direct investment may be a means of pre-empting entry into an industry by other firms (A. Smith, 1994) or of increasing market concentration, so that the benefits of co-operation are attained without collusion. Hence, common ownership may result from strategic decisions by the firm rather than location-specific advantages.

In its extreme form, vertical integration involves common ownership of the travel agencies in which tourists purchase their holidays, the tour operators which assemble them, the airline or other form of transport in which tourists travel and the hotels in which they stay. Tourists may also eat and drink imported food and beverages, so that the destination receives very little foreign currency from their stay. Conversely, in cases where the destination's airline and locally owned accommodation are used and local products consumed, both the value added in the destination and the foreign currency earnings accruing to it are higher. Thus, different types of integration affect the international distribution of returns from tourism. The case of Kenya exemplifies the possible distribution of expenditure on package holidays between the destination and origin countries. Taking a fourteen-day package holiday in a beach location in April and December of 1990, Kenya's share of UK tourists' expenditure was only around 38 per cent if the tourists travelled to and from the country in a UK airline (Sinclair, 1991b). Losses were incurred in the form of payment for items imported for tourist consumption – approximately 30 per cent of expenditure on drinks and less than 20 per cent of expenditure on food. Kenya's share of spending on a fourteen-day beach and safari holiday was around 66 per cent. The higher figure is partly due to the fact that tourists on beach and safari holidays usually travelled by air between Nairobi, the usual departure point for safaris, and coastal airports and were required by the Kenyan government to use the national airline. Kenya's share was considerably higher, approximating 80 per cent, if tourists travelled to and from the country using Kenya Airways. Hence, tour operators' use of origin- or destination-based airlines is a particularly important determinant of the destination country's share of tourism revenue. If tour operators and airlines are commonly owned, as is often the case, operators make preferential use of their own airlines and the destination's share of tourist expenditure declines.

135

Like vertical integration, horizontal integration has been widespread within the tourism industry and acquisitions, mergers and takeovers have increased the degree of industrial concentration so that a smaller number of firms control a larger percentage of total supply. The higher levels of common ownership of UK travel agencies and tour operators, discussed in Chapter 4, are examples of increasing concentration. Firms also integrate horizontally across countries, as in the case of the Franco-Belgian company Wagon Lits' investment in the Spanish travel agent Viajes Ecuador or the German operator TUI's majority holding and the Holland International Travel Group's minority holding in Ultramar Express, the Spanish travel agent and tour operator (Bote Gómez and Sinclair, 1991). Firms may integrate horizontally with the objectives of increasing efficiency by means of scale economies and of increasing market power, enabling firms, potentially, to raise prices and profitability. Integration may be a means of increasing the firm's growth and market share, increasing barriers to entry, increasing the market valuation of the firm and obtaining easier access to finance (Prais, 1976). It may also take place in order to decrease uncertainty, providing the firm with greater control over its environment (Newbould, 1970; Aaronovitch and Sawyer, 1975). Horizontal integration can take forms other than common ownership including joint reservations arrangements between airlines, such as BA's code-sharing agreements with US Air, or management contracts or franchising agreements between hotels. Examples of the latter include Holiday Inn's franchise agreement and management contracts with hotels in Thailand and the Thai Dusit Thani Group of hotels' franchise agreements with domestic hotels, management contracts with hotels in Indonesia and planned hotel and tourism development projects in Laos and Vietnam (Sinclair and Vokes, 1992).

Within the international hotel sector, multinational corporations (MNCs) have engaged in both horizontal integration in the form of foreign direct investment, leasing, management contracts, franchising and marketing agreements and vertical integration between hotels, airlines, tour operators and travel agencies. For example, approximately 78 per cent of major hotels along the Kenyan coastline and around 66 per cent of those in Nairobi and National Parks and Reserves have had some foreign investment, although less than 20 per cent have been subject to total foreign ownership (Sinclair et al., 1992). Equity holdings have been acquired, for instance, by the tour operators African Safari Club, Hayes and Jarvis, Universal Safari Tours, Kuoni, Polmans, TUI, Franco Rosso and I Grandi Viaggi, as well as by BA and Lufthansa airlines. A number of hotels have been managed by foreign hotel chains, and franchising contracts with such groups as Holiday Inn and Hilton have occurred.

Some MNCs have had different types of integration with a given host country firm, for example a minority equity holding in a hotel combined with a management contract or franchise; majority equity holding and shared management with other MNCs; or leasing combined with equity holding, management or marketing arrangements (Dunning and McQueen, 1982b). In the early 1980s, non-equity integration was characteristic of many French, Japanese and American firms' relationships with hotels in developing countries, the MNCs supplying technology, management and marketing expertise. Ownership was more common by firms based in other West European countries. Equity investment combined with franchising was fairly common in Asia and equity and leasing was characteristic of hotels in industrialized countries and the Caribbean. Hotels in some West African countries, where there is government opposition to majority ownership, engaged in management contracts with MNCs, as did hotels in some Middle East countries which had finance capital but lacked a skilled labour force. Franchising and marketing agreements tended to be more common in destinations with greater local expertise, notably Brazil and Mexico. The significant heterogeneity of relationships between MNCs and host country firms in the international hotel sector demonstrates the importance of cross-country integration but the inappropriateness of global generalizations about their nature.

Although, it might appear that foreign ownership of tourism firms is disadvantageous, a combination of foreign and domestic ownership has the advantage of spreading the risks associated with running tourism enterprises. It also provides the foreign participants, such as tour operators or airlines, with a strong incentive to maintain or increase the demand for tourism in the destination and can supply additional business expertise and increase investment in the area, as in the case of Australia (Dwyer and Forsythe, 1994). On the other hand, governments may spend large amounts on infrastructure for the tourism sector, payments of profits and/or wages to foreign nationals are sometimes high and some foreign firms pass little expertise on to the recipient country. For instance, in hotels in the British Virgin Islands and Grand Cayman, at the beginning of the 1970s, at least 43 per cent of wages and salaries were estimated as paid to expatriates who predominantly filled the managerial and professional positions (Bryden, 1973). At the same time, many Caribbean governments also offered highly favourable fiscal incentives for hotel development, including the provision of credit on very advantageous terms, tax concessions and considerable infrastructure provision.

Contractual arrangements between tourism firms in different countries vary enormously from case to case, providing some countries with significantly more benefits than others (Dunning and McQueen, 1982a). Firms in many developing countries have little knowledge of the contractual terms which prevail elsewhere. In contrast, firms from developed countries have notable informational and first-mover negotiating advantages, as in the case of tour operators' contracts with hoteliers in developing countries. Over the longer term, a country's ability to move on to higher growth paths depends, to a great extent, on its ability to acquire additional knowledge and foreign direct investment and contractual arrangements between firms provide possible means of doing so. However, the wide variations in the types and extent of foreign participation and the fact that foreign participants not only provide but also acquire local knowledge (Daneshkhu, 1996) mean that the outcome for the host country may be growth-enhancing or immiserizing (de Mello and Sinclair, 1995).

Strategic tourism policy

An important difference between imperfect competition in the domestic and international context is that, in the latter case, governments have an incentive to implement strategic trade policies (Krugman, 1989a). Strategic policies aim to help firms of their own nationality to achieve higher export earnings and/or to decrease outflows of foreign currency in payment for goods and services produced abroad. They include strategic commercial policy in the form of export credits or subsidies or import tariffs. For example, in the travel context, subsidies may be granted to state airlines or taxes imposed on airport departures. Devaluation of the country's exchange rate and measures to restrain inflation increase the price competitiveness of tourism in the country. Strategic industrial policy includes, in the case of tourism, payments to domestic businesses, such as capital grants or low-interest loans for improvements in hotel and guesthouse accommodation and state aid for research and development, for example, in information technology relevant to travel agents, tour operators, car hire firms, airlines and hotels. The government can also threaten to alter the accessibility conditions or route allocation to foreign airlines. For example, competition on the high density routes between high income countries is usually precluded by governments' allocation of favourable arrival and destination slots to their own national carriers and imposition of limitations on the numbers of flight arrivals by airlines from other countries. In some circumstances the government may not have to implement its proposed policy, as a well-publicized threat to restrict landing and take-off slots may be sufficient to deter foreign competitors from competing in the market. Thus, the range of strategies which are available to domestic firms is widened by the government's actual or potential policies, so long as such policies are known and credible.

In practice it is difficult for developing countries to gain first-mover advantages, as competitive initiatives usually come from developed countries, which have acquired greater expertise in establishing competitive prices and supply. Relatively small country size may also impede a government's ability to engage in strategic policy formulation (Krugman, 1989b) and governments may be subject to lobbying pressures from foreign investors or governments or may not possess sufficient information about the operation of specific sectors to intervene effectively (Alam, 1995). Developing countries' attempts to restrict the numbers of incoming flights from particular origin countries is a case in point, as tour operators sometimes successfully evade the restrictions. For example, tour operators avoided the Kenyan government's attempts to limit air traffic between London and Nairobi to scheduled flights by routing charter flights via Italy (Sinclair et al., 1992). In the case of Kenya, effective state control was maintained only over domestic flights, which were limited to the national airline, in which KLM subsequently obtained an equity holding.

In deciding upon whether to propose or implement a tourism strategic policy, the government must predict other governments' responses to the policies under consideration. The interrelationships between strategies of governments, like those of firms, may be analysed by means of game theory, examined in Chapter 5. For example, the case of potential subsidies to firms is analogous to the well-known case in game theory of the prisoner's dilemma. Although all countries might achieve a pay-off in the form of higher net income in the absence of subsidies, each would be worse off if others were to subsidize their producers. It is clear that a range of outcomes may occur if governments engage in strategic policies to benefit their own producers, as well as to deter potential competition, and if firms formulate their strategies in the context of their governments' policies and other governments' and firms' likely responses to them. As shown in Chapter 5, the advantage of using game theory to examine optimal strategic tourism policy is that it makes explicit the range of possible courses of action and outcomes and the advantages and disadvantages which are associated with each. Unlike the comparative statics approach of comparing pre- and post-policy positions, it can shed light on the dynamic process in which both firms and governments are active players at the global level.

EFFECTS OF INTERNATIONAL TOURISM

Income and employment multiplier effects

The first part of this chapter showed that a variety of theories may explain different aspects of international tourism. The second part will consider some of the effects which tourism can bring about, starting with changes in income and employment, which are of concern to private individuals and firms as well as the public sector. Tourism creates income and employment directly in the sectors in which expenditure or tourism-related investment takes place and also induces further increases throughout the economy as the recipients of rises in income spend a proportion of them. Income and employment generation result not only from expenditure by foreign tourists, along with the associated increases in private investment and public expenditure, but also from domestic tourist expenditure, which often exceeds that by foreign tourists. It is useful to consider the effects of both types of expenditure within an international context in order to take explicit account of the leakages abroad in such forms as payments for tourism-associated imports

139

and remittances of income, profits and dividends, which tend to be important in small, open economies.

The extent to which changes in tourism demand are likely to affect income and employment depends, in part, upon the existence of the appropriate surplus resources within the economy. For example, airport capacity must be sufficient to meet additional demand by incoming and departing tourists, hotels and guesthouses must have less than full occupancy rates so that more tourists can be accommodated and labour with the appropriate skills must be available to provide the additional services required. If surplus capacity is not available, as in periods of high seasonal demand, additional demand results in rising prices and wages rather than increases in real income. Monetarists such as Friedman (1968) argue that if the economy has stabilized at the 'natural rate of unemployment', subsequently re-conceptualized as the non-accelerating inflation rate of unemployment (NAIRU), increases in demand give rise to an increasing rate of inflation rather than a rise in real output, even in the context of spare capacity. New classical economists (for example, Lucas, 1972; Sargent and Wallace, 1976) add the assumption that people use all available information when making forecasts (rational expectations) so that short-run changes in real output occur only when they are mistaken. Hence, it is necessary to examine whether there is surplus capacity in the economy in order to assess the potential for increases in demand to raise output, and the way in which consumers and business people formulate their expectations about the future in order to determine whether increases in demand result in higher output or inflation. The assumption that spare capacity is available and is utilized to produce more output in response to an increase in demand is typically Keynesian and provides the basis for the use of the multiplier methodology to estimate the value of income and employment which is generated by tourism. If surplus capacity is not available, the use of the methodology is likely to provide spurious results. Wanhill (1988) attempted to take account of this problem by introducing capacity constraints into his multiplier model.

Two models which can be used to estimate income and employment generation resulting from tourism are the Keynesian multiplier model and the input–output model. The models will not be discussed in detail here as they have already been the subject of much attention (for example, Archer, 1973, 1977a, 1989; Sinclair and Sutcliffe, 1988a, 1988b, 1989a; Johnson and Thomas, 1990; Fletcher and Archer, 1991). Both approaches permit the calculation of the value of the multiplier, which is the ratio between the value of income or employment generated and the initial change in tourist spending or tourism-related investment. The Keynesian income multiplier model involves the use of a range of equations to estimate the aggregate multiplier values associated with different types of tourist expenditure or investment and the alternative definitions of income which can be measured: gross domestic product (GDP) or gross national product (GNP) at market prices or factor cost or disposable income (Sinclair and Sutcliffe, 1988a). In the case of tourist spending in the Spanish province of Malaga, for example, it was estimated that the multiplier value for GNP at factor cost was 0.72 and that for disposable income was 0.54. This implies that for every 1,000 pesetas of tourist expenditure, provincial GNP increased by approximately 720 pesetas and the disposable income of the local population rose by around 540 pesetas. Thus, the calculated multiplier and income generation values also vary according to the definition of income which is being measured, although few authors have provided an explicit definition of the type of income which they are measuring. Multiplier values for the change in GNP or GDP are likely to be significantly higher than those for disposable income. The calculation of GNP or GDP multiplier values alone may be misleading because it is disposable income which remains in the hands of the local population after remittances, income tax and national insurance payments have occurred. Tourist expenditure and investment also vary over the tourism season and time, causing different related short- and long-run multiplier and income generation values (Sinclair and Sutcliffe, 1989a).

Like the Keynesian income multiplier model, the input–output model permits the calculation of the aggregate multiplier values for different definitions of income. However, it also provides estimates of the multiplier values for different sectors of the economy, such as food and drink provision, electrical equipment, textiles and cleaning materials. It, therefore, demonstrates the relative importance of the interrelationships between the different sectors at a given point of time. This model has been applied to a wide range of countries including Antigua (Pollard, 1976), the Bahamas and Bermuda (Archer, 1977b, 1995), Hong Kong (Lin and Sung, 1983), Korea (Song and Ahn, 1983), the Philippines (Delos Santos et al., 1983), Palau, Western Samoa and the Solomon Islands (Fletcher, 1986a, 1986b, 1987), Mauritius (Wanhill, 1988), Singapore (Heng and Low, 1990; Khan et al., 1990) and the Seychelles (Archer and Fletcher, 1996).

Accurate calculation of both Keynesian and input–output multiplier values involves, in particular, the careful estimation of the first round leakages from tourism demand (Sinclair and Sutcliffe, 1978). These may be relatively high, resulting in low multiplier values. In some Caribbean countries, for example, the import content of beverages and cigarettes has been as high as 69 per cent and that of food 62 per cent (Cazes, 1972). Considering countries in other parts of the world, the import content of beverages and food in Fiji has been estimated as 45 per cent and 56 per cent respectively (Varley, 1978), while the overall import content of tourism expenditure in The Gambia was 55 per cent (Farver, 1984). In contrast, the import content of beverages in Kenya was approximately 35 per cent and that of food 10 per cent (Sinclair, 1991b) and there is scope for decreasing the import content in other countries, as illustrated by the case of food production in Lombok, Indonesia (Telfer and Wall, 1996). The corresponding values for most developed countries, for example Spain (Sinclair and Sutcliffe, 1988b), are also relatively low. Hence, the associated multiplier and income generation values vary considerably between destinations.

It might, at first sight, be supposed that high multiplier values are more beneficial to the local economy than low values. However, it is necessary to consider both the multiplier value and the aggregate value of the income or employment generation as it is possible for a high multiplier value to accompany a low value of total income generation owing, in part, to a low initial level and change in demand. For example, the multiplier values associated with small-scale establishments may be high owing to the high local content of tourists' expenditure on them but the value of expenditure on them, both in per capita and in aggregate terms, tends to be low, so that the magnitude of income and employment generated is low. In contrast, the multiplier value associated with expenditure on higher category hotels may be low because of the higher import content of tourists' expenditure, but total income and employment generation is usually relatively high, owing to the high value of tourist expenditure on them.

Input–output multiplier studies can provide a large amount of information about inter-sectoral income and employment generation but, like Keynesian income multiplier models, they neglect the personal income distribution effects which result from the change in tourism demand. Each type of tourist expenditure is associated with different distributional repercussions and although the aggregate welfare gains may be significant, there may be adverse effects on particular individuals and groups, notably those who do not own the land used in the service sector (Copeland, 1991). Thus, a promotional strategy which favours, for example, luxury coastal tourism is likely to provide additional income for property developers but often results in negative externalities such as marine pollution which may adversely affect the livelihoods of poorer fisherpeople. Moreover, local people who are displaced from the areas of tourism growth may be inadequately compensated for their loss of land and therefore experience unfavourable wealth effects, as in the case of Santa Cruz in Mexico (Long, 1991). Such effects should be considered alongside the more readily quantifiable income and employment multiplier effect but are all too often ignored. If the probable costs and benefits, both economic and non-economic, of different types of tourism were made explicit, a choice between them could be made in the context not only of their estimated income and employment generation but also of their wider social and distributional consequences, as well as their environmental impacts. Some of the structural changes which are associated with tourism are considered in the following section and the environmental effects are examined in Chapters 7 and 8.

Structural and migratory repercussions

Studies using the multiplier methodology take account of the short-run effects of tourism on income and employment but the models are not appropriate for examining its longer-term repercussions. These affect both the structure and the spatial distribution of economic activity. Taking, first, the structural changes, one of the most obvious effects of tourism expansion is the rise in the importance of the service sector within the economy. In developing countries this is usually accompanied by a fall in the agricultural sector's share of gross national product, while many developed countries experience a process of deindustrialization in addition to declining primary sector production, to which the growth of tourism may contribute (Copeland, 1991; Adams and Parmenter, 1995). Surprisingly, little attention has been paid to the growth of the service sector in either context. The development literature has concentrated, in the 1950s and 1960s, on the growth of manufacturing via the transfer of supposedly surplus, low-wage labour from agriculture to industry. From the late 1960s onwards, relatively more attention was paid to agriculture as the problems of poverty and unemployment in the cities demonstrated the inability of the manufacturing sector to assimilate the migrants from the land. It was not until the late 1980s that the need to examine the service sector was generally recognized (UNCTAD, 1988), although tourism was rarely acknowledged in either developing or developed countries.

Empirical evidence about the role of tourism within the structural transformation of economies over the long run is limited but indicates that the common conception of the development process as a transition from an agriculture-based economy to manufacturing and thence to services may be misguided. In Spain, for example, it was the large-scale receipts from foreign tourism, as well as remittances from Spanish nationals working abroad, often in tourism-related activities, which underpinned the country's industrialization process (Bote Gómez, 1990, 1993). Much of the foreign capital which flowed into the country was channelled into the tourism sector, particularly purchases of property which constituted 20 per cent of total foreign investment in 1964 and varied between 59 per cent in 1974 and 10 per cent in 1991 (Sinclair and Bote Gómez, 1996). The changes in land values which resulted from planning permission for construction, particularly for tourist accommodation in coastal areas, along with the profits made from the construction and sale of hotels, flats and villas, resulted in large increases in the wealth of national and foreign property developers. Interestingly, Spain has now become a significant investor in tourism expansion abroad, as is illustrated by its growing participation in the hotel sector in Cuba.

The case of tourism in Spain clearly demonstrates the rapid growth and spatial concentration which can characterize tourism demand and supply. Whereas in 1951, 1.3 million people came to Spain, by 1990 over 52 million arrived from abroad, in addition to more than 13 million domestic tourists. The vast majority of tourism demand occurs in the Mediterranean coastline, Balearic and Canary Islands, which accounted for three-quarters of demand for and supply of officially registered bed-places in 1990. Foreign tourists, in particular, have strong preferences for beach rather than rural and city tourism and the expansion of tourism in coastal areas created an important migratory movement from villages in the interior to meet the rising demand for cheap labour to work in tourism activities. It is only in isolated cases that rural tourism development has succeeded in decreasing the population out-flow. The examples of tourism in La Vera in the west of Spain and Taramundi in the north illustrate the small-scale but significant effects that local tourism development programmes can bring about (Bote Gómez, 1988, 1990; Bote Gómez and Sinclair, 1996).

In some contexts, the expansion of leisure and tourism in rural areas has created considerable income and employment and increased the spatial dispersion of tourism, such as in Britain during the 1990s (D. Diamond and Richardson, 1996). In Egypt, in contrast, the growth of tourism in the interior has been constrained by well-publicized violence against a small number of tourists. Large-scale construction, with probable adverse environmental consequences, is occurring in the coastal areas of the north and the Red Sea, resulting in migration from the cities and towns to these relatively unpopulated areas (The *Economist*, 1996). Although many of the newly created jobs have been taken by men, some Egyptian women have been able to obtain supervisory roles in large modern hotels. Thus, although the major gains from tourism have clearly gone to property speculators and construction companies, tourism has also altered the distribution of employment and income at the most basic level of individuals and households. Some of these effects will be discussed in the following section.

Gender structuring of tourism employment

Tourism brings about changes in the pattern of employment, as well as in the distribution of income. Relatively little attention has been paid to labour market analysis of tourism and the following discussion focuses on research which has been undertaken on the gender structuring of tourism employment, while acknowledging that considerable further examination of labour market theories and tourism employment is required within the literature. Studies of employment in tourism have indicated the ways in which it, like other sectors of the economy, is structured by gender (Bagguley, 1990; Kinnaird et al., 1994; Adkins, 1995; Swain, 1995; Sinclair, 1991c, 1997b). In the UK in 1995, for example, 76 per cent of jobs in the transport sector were filled by men while 62 per cent of those in accommodation and catering were undertaken by women (Purcell, 1997), indicating horizontal segmentation of work. Vertical segmentation also occurs as most top jobs are carried out by men (Guerrier, 1986; Hicks, 1990; Purcell, 1997). Since women's labour supply is concentrated in a limited number of occupations which are not characterized by a tradition of trade union activity, the wages which they are paid tend to be relatively low. The majority of seasonal, part-time and low-paid tourism work in many areas is undertaken by women (Breathnach *et al.*, 1994; Hennessy, 1994), in combination with their childcare and domestic duties.

In developing countries and in rural areas in some intermediate income countries, many of the tasks which involve direct contact with tourists are carried out by men. This is exemplified in the case of Turkish Cyprus. Local women have traditionally undertaken behind-the-scenes work such as cleaning and bed-making, which are akin to their domestic roles, although women from Eastern Europe are employed as croupiers in casinos (Scott, 1997). The allocation of domestically related duties to women has also occurred in a range of other areas, including Greece (Castelberg-Koulma, 1991; Leontidou, 1994) and the Caribbean (Momsen, 1994). In some countries, gender norms are less restrictive so that women also work in tourism businesses, for example in Western Samoa (Fairbairn-Dunlop, 1994), but traditional expectations concerning the roles which it is appropriate for married women to undertake often constrain their activities, as in tourism-related work in Bali (Long and Kindon, 1997).

Some insights into the gender structuring of tourism employment may be obtained by examining the demand for and supply of labour in relation to prevailing societal norms (Sinclair, 1997c). In countries which are characterized by strong traditions concerning the roles which are deemed suitable for men and women, employers hire (demand) male workers to fill some posts and women to fill others, while men and women tend only to apply for (supply their labour in) positions which are viewed as appropriate for their gender. In other countries in which capitalist development has altered prevailing norms, employers are more indifferent concerning the gender of those who undertake different jobs, although applicants remain subject to modified expectations concerning gender roles. Hence labour demand and supply not only vary in relation to quantifiable variables such as wage rates, taken into account by mainstream labour economists, but also depend on norms and expectations which are not easily measured.

It is interesting to consider how international tourism can bring about changes in the structure of employment. In Northern Cyprus, for example, accommodation has traditionally consisted of small, family-based guesthouses used by domestic tourists (J. Scott, 1997). Social and financial transactions with tourists are usually undertaken by male household members while women are responsible for cleaning and other household work. However, an influx of foreign tourists has been accompanied by the construction of large hotels in which many younger and more educated Cypriot women have obtained employment. Gender norms in both the demand for and supply of labour in the new hotels differ from those prevailing in the guesthouse sector, providing more opportunities for local women. Therefore, it may be necessary to examine labour demand and supply in different sub-sectors of tourism supply.

Although tourism can provide local women with higher levels of income and greater independence within the household, it can also result in considerable problems (Chant, 1997). Prostitution tourism is an obvious example. This has provided work and income for many women, as well as men and children, often from poor rural areas, and has sometimes received implicit support from the state (W. Lee, 1991). However, it has involved harsh conditions for many of those involved. Campaigns to discourage sex tours have had some success in deterring tourists from travelling to countries known for their high levels of prostitution. However, they have also had the effect of encouraging 'trade' in women, as gangster syndicates have recruited poor women from developing countries to work as prostitutes in such wealthy countries as Japan (Muroi and Sasaki, 1997). The intermediaries extract high profits whereas the women themselves are usually unaware, on leaving their homes, that they have been recruited to work as prostitutes rather than entertainers, receiving a very small share of the returns and often of being unable to escape from their new destination.

The examples of tourism in Northern Cyprus and of prostitution tourism illustrate the role which both cultural norms and economic variables play in determining the nature of employment. International tourism results in the interaction of different, historically based and gendered cultural systems which are modified but not standardized by tourist inflows. Hence, countries' specific cultural endowments, as well as their particular combination of land, labour and capital, determine the patterns of specialization, trade and employment in tourism and other activities.

International tourism and instability in export earnings

Tourism has been one of the main sources of foreign currency in many countries but is also perceived as a sector which is particularly prone to instability. In Kenya, for example, earnings from tourism were more important in absolute terms than those from any other export in the late 1980s but real dollar earnings, both total and per tourist, were lower during the 1980s than during most of the 1970s (Sinclair, 1990, 1992b). Tourist expenditure and investment in tourism services vary between different locations and time periods owing to such causes as seasonality, one-off sporting or cultural events which raise demand for a short period or political instability which depresses it. The associated multiplier effects on income and employment are negative where tourism demand decreases or, assuming surplus capacity, positive if it rises. The changed levels of income and employment persist only if the change in tourism expenditure persists; if not, they occur on a one-off basis, after which income and employment return to their initial levels. Thus, tourism-related income, employment, foreign currency generation and the welfare of local inhabitants can vary considerably over time.

146

Instability in tourism receipts can have a range of adverse repercussions on destination countries (Rao, 1986). In addition to the simple multiplier effects on income and employment, a decrease in the rate of growth of income from tourism and other sectors of the economy can induce a fall in investment. Investment also depends on potential investors' expectations, which may be affected adversely by instability in foreign currency earnings and uncertainty about its cost and availability. A further repercussion is that purchases of capital goods imports may be deterred which, in turn, will cause instability in the fiscal yield from import and export taxes as well as from income and indirect taxes, thereby possibly constraining government spending on infrastructure and services. If such effects occur, they are likely to lower the rate of growth, particularly in developing countries which have low levels of foreign currency reserves.

A somewhat different view of the effects of export earnings instability, put forward by Knudsen and Parnes (1975), is based on Friedman's (1957) permanent income hypothesis of consumption. According to their theory, increases in foreign currency receipts only have an expansionary effect on consumption if they are viewed as a permanent increase in income, as Friedman argued that consumption depends on permanent rather than transitory income. Knudsen and Parnes hypothesized that if earnings from exports are unstable, increases in them are not likely to be perceived as permanent but as transitory income and are saved rather than consumed. The lower propensity to consume and higher propensity to save do, however, provide funds which can be channelled into higher investment, resulting in higher growth. In this way, export earnings instability is said to have an advantageous, rather than an adverse or neutral effect on the economy, the implication being that instability in tourism demand should not be discouraged. In contrast, new classical macroeconomists posit that changes in demand have no real effect unless they are unanticipated and have not been discounted in advance.

There is little empirical research on the measurement or effects of tourism earnings instability, partly owing to the problem of selecting an appropriate definition of instability but also to the fact that this issue is related to the wider issue of the appropriate specification of consumption, savings and investment functions (Deaton, 1992). A study of a range of developed, intermediate income and developing countries, using different measures of instability (including the mean or standard deviation of absolute deviations from the trend in export earnings, estimated using multiple regression analysis or a moving average), found tourism earnings to be relatively unstable (Sinclair and Tsegaye, 1990). The results indicated that, for most of the countries considered, variations in tourism receipts were not accompanied by offsetting variations in receipts from exports of goods, so that the net effect appeared to be an increase in instability. Those countries in which instability has adverse repercussions could implement policies to diversify the mix of tourist nationalities or types of tourism so as to stabilize net receipts or to reduce dependency on tourism by developing other forms of economic activity. The desired mix varies according to the preferences of policy-makers concerning the trade-off between higher values and higher variability in receipts and their perception of future trading conditions in all sectors of the economy. Within tourism, alternative combinations of nationalities or tourism types are associated with differences in the values of receipts (the 'return' from tourism) and in their variability over time (the associated 'risk'). Portfolio analysis, a technique used in the financial analysis of the risks and returns associated with investors' portfolios of stocks and shares, can be used to estimate the combination of nationalities or tourism types which would constitute policy-makers' preferred mix or tourism portfolio (Board *et al.*, 1987); appropriate incentives can then be provided.

International tourism and economic growth

Debates about the effects of export earnings instability have been particularly concerned with the positive or negative consequences of instability for growth. Little attention has, so far, been paid to the topic of growth as the discussion in this chapter has been conducted within a static context or within a comparative static framework in which an initial position is compared with a subsequent outcome, for example, the levels of income and employment before and after a change in tourist spending. Comparative advantage was also examined within its traditional static context of given technology and resource endowments, giving rise to a specific pattern of international specialization and trade. However, it is useful to consider further the possible effects which international tourism can have on economic growth. One reason for doing so is that countries can become entrenched in specific patterns of production and trade, which may affect their long-run growth adversely or beneficially. For example, developed countries tend to specialize in producing and exporting manufactured goods with high income elasticities of demand, while developing countries produce and export primary commodities with low income elasticities. International tourism may reinforce this type of specialization as developed countries supply the high expenditure, high income generation, high growth components of tourism while developing countries tend to supply environmental resources and accommodation and local transport, which may be subject to foreign ownership or control.

The growth differences between developed and developing countries may become self-perpetuating as the former take advantage of economies of scale which are supported by trade (Rivera-Batiz and Romer, 1991). Developing countries are usually unable to benefit from ongoing innovation and, consequently, convergence of growth rates between countries fails to occur. The obvious implication for countries on low growth paths is to alter their comparative advantage by specializing in high-growth exports, including tourism products with a high-income elasticity of demand. Thus, tourism provides a possible means for lower-income countries to escape from the low product quality, low expenditure and low-income pattern which generally constrains their development. It is, therefore, useful to consider comparative advantage in the context of economic growth theory in order to examine whether and how it can change over time, i.e. dynamic comparative advantage. A number of theories of economic growth will now be discussed. The ways in which countries may change their comparative advantage and increase tourism's contribution to growth will be considered subsequently. The discussion will take the form of an overview of relevant ideas and issues as analysis of the relationship between tourism and growth has yet to be undertaken.

New growth theories

Economic growth has been a key area of attention within economics in recent years and new growth theories have modified much traditional thinking. The neoclassical view which dominated much of the literature until the early 1980s (Solow, 1956) argued that growth depended upon the supply of labour and capital, with any residual growth being determined by exogenous technological change (independent of the economic system). The economic context was assumed to be one of perfect competition, with labour and capital being subject to decreasing returns. Thus, each additional unit of labour or capital would make a smaller contribution to output than the previous unit, in the context of given quantities of other units of production. The main competing theory was put forward by Harrod (1939) and Domar (1946, 1947), who argued that growth was given by the ratio of the savings rate to the capital–output ratio; that is the input of capital required to achieve a given level of output, the required capital input being a constant proportion of output with a fixed capital–output ratio. If the actual growth rate exceeds business people's expected growth rate, they increase their investment and the actual growth rate increases further. The opposite would occur if the expected growth rate exceeded the actual outcome. Thus, within this approach, expectations play a key role in determining growth.

Economic growth is determined endogenously (within the economic system) according to new theories of growth (for example, Grossman and Helpman, 1994; Romer, 1994; the review by Van der Ploeg and Tang, 1994). Capital is defined not only as physical capital in the form of equipment and machines but also as public infrastructure and as human capital, for example in the form of skilled labour. Economic growth can arise from investment in the broad definition of capital, including investment in knowledge. It is assumed that there can be substitutability between capital and labour and between different forms of capital, and that there are constant returns to the broad definition of capital so that there is no incentive to decrease investment in capital. Firms can operate under imperfect competition, thereby reaping supernormal profits in the long run from their investments.

However, any given firm is unlikely to be able to appropriate all of the benefits from its own investment expenditure so that investment by one firm has positive external effects on other firms at the national and/or international levels as information is diffused about new products or methods.

New growth theories consider that growth can result from the accumulation of human capital, involving education and training of present and future workers (Lucas, 1988). Although the case for education and training to raise the rate of growth in tourism and hence the economy as a whole may appear obvious, it is only relatively recently, in the UK for example, that the need for training in tourism has been widely recognized and increased provision made. Moreover, there is still some debate about the form of training which is most appropriate; consequently its quality varies considerably both within and between countries. 'Learning by doing' is a second form of knowledge accumulation (Arrow, 1962; Romer, 1986; Young, 1991) and within the management literature there has been considerable debate about 'empowering' workers to make an increased contribution to improving firms' performance. In the tourism industry, some hoteliers make arrangements for members of staff, including managers, to work in other hotels which are considered to be of a comparable or superior standard. In some cases this involves sending staff to other countries, so that the ultimate objective is a cross-border transfer of knowledge and skills.

Research and development has been proposed as a third determinant of growth (Grossman and Helpman, 1990a, 1990b, 1991; Aghion and Howitt, 1992) and can increase the variety and quality of products supplied. Like learning by doing, research and development can increase the general availability of knowledge beyond that of the individual firm undertaking the development. This may occur relatively quickly, as new tourism products are marketed, or more slowly if innovating firms attempt to appropriate the returns from the innovation, as in the case of computer reservation systems for holiday bookings.

Growth can also be facilitated by the provision of public infrastructure either in material form such as roads for local residents and tourists or in non-material form, for example, health care (Barro, 1990; Barro and Sala-i-Martin, 1992). The view that public provision of infrastructure can further the growth of the private sector contrasts with much of the thinking that was dominant during the monetarist era, when government spending was thought to crowd out private consumption and investment via increases in prices and interest rates. New growth theorists acknowledge that public sector provision may be financed by distortionary taxes which can be inequitable but show that the net effect on growth can be positive rather than negative. Within tourism, infrastructure is seen as facilitating the sector's growth and is referred to as the secondary tourism resource base.

The role of initial factor endowments in the new growth theory depends upon the degree of international spillovers of knowledge. For example, if firms invest in a new product or process and knowledge about their investment is available to domestic but not foreign firms, the positive externalities of the investment are retained within national boundaries. In this respect, the country which has the initial advantage in knowledge and investment retains a comparative advantage in products which incorporate this type of knowledge, irrespective of its factor endowments. If, on the other hand, there are no restrictions on access to knowledge, no country can gain a comparative advantage in supplying the product so that comparative advantage is, instead, determined by countries' factor endowments. In the case of the accommodation component of tourism, for example, hoteliers in some developing countries may be unaware of innovations in accommodation and service provision in wealthy countries, such as improved computing equipment and software. Even if they are aware of such innovations, many firms in developing countries are unable to take advantage of them owing to a shortage of investment funds and experience. Thus, what is important is the extent and speed of intra- and inter-country knowledge generation and transmission and firms' ability to take advantage of the knowledge which they obtain.

Tourism policy implications

It is clear that in contrast to traditional neoclassical theory, new growth theory provides a possible role for government (van der Ploeg and Tang, 1994), although government intervention does not always appear beneficial (Barro, 1991). The measures which the government can undertake include direct expenditure on or measures to encourage investment in a broad range of human and physical capital. In the case of tourism, the natural and built environment should also be taken into account. Such investment, along with the provision of a complementary institutional framework, may also succeed in shifting the country's comparative advantage towards targeted sectors. Improved provision of roads, airports, health care and disease eradication campaigns in developing country destinations increase both tourism demand and supply and may increase growth. Tourists gain increased and/or improved accessibility to different areas within destinations and are confident that their health needs will be met, while private sector suppliers are encouraged to provide more accommodation and supporting facilities within reasonable proximity of the new infrastructure; additional airport provision in Cambodia is a case in point. Investment in natural resources as well as physical and human capital should also be undertaken if growth is to be sustainable, as pointed out in the environmental economics literature considered in Chapters 7 and 8. Legislative measures can attempt to provide an institutional framework which facilitates sustainable growth and commercial, fiscal and industrial policy measures can be used to stimulate investment, such as in the provision or modernization of tourist accommodation.

Where domestically generated savings are insufficient to cover investment requirements, governments have sometimes encouraged additional investment by deregulating foreign direct investment in tourism enterprises, as in the case of Indonesia's Fourth Five Year Plan, Repelita IV, 1983/4–1988/9. The potential for international collaboration to shift the recipient country on to a higher growth path is, however, dependent upon an optimal degree of knowledge transfer to the host economy, as a low level has a small effect on growth while an excessive level crowds out local firms (de Mello and Sinclair, 1995). Thus, governments of some host economies have imposed conditions concerning the degree and nature of technology transfers by foreign investors which are required in return for permission to invest in the country (Chakwin and Hamid, 1996) but have paid little attention to this issue in the context of tourism investment.

Higher standards of education and training, promoted and/or financed by governments or international organizations, are a further means of supporting ongoing growth. The case for investment in both human and physical capital has been acknowledged by governments and international organizations. For instance, centres for training in tourism in Bandung and Bali in Indonesia received aid, respectively, from Switzerland, the International Labour Organization (ILO) and the United Nations Development Programme (UNDP) and the State University of Udayama provided Diploma and Master level courses in tourism from the mid-1980s (Pack and Sinclair, 1995a). Government institutions have supplied tourism training in India, aided by the ILO (Pack and Sinclair, 1995b), as in many other countries. Investment in physical capital for tourism has been undertaken or supported financially by national governments, and international organizations including the Asian Development Bank (ADB), World Bank, World Tourism Organization (WTO) and United Nations have supplied financial and/or technical assistance for infrastructure provision, as well as for more general tourism development programmes (for example, ADB, 1995). 'Regional' groups such as the European Union have also considered a variety of measures to support tourism and increase welfare (Commission of the European Communities, 1994, 1995a, 1995b, 1996).

If appropriate measures are implemented in both tourism and other sectors of the economy, they may assist not only in promoting endogenous growth but also in moving the economy on to a higher growth path by shifting a country's comparative advantage away from products with a low income elasticity of demand. Examples include the Singapore government's encouragement of increases in investment, education, training and wages, which stimulated production of more technology-intensive, higher-priced goods (Chadha, 1991; Tan, 1992) and the South Korean government's encouragement of conglomerates, which proved a favourable institutional structure for investment and growth during the 1980s (Mody, 1990). Comparative advantage, thus, becomes dynamic and subject to strategic policy manipulation.

Hence, identification of the conditions under which tourism could constitute a locus of specialization and source of higher growth in particular countries, as well as increasing welfare in them, is clearly important.

CONCLUSIONS

This chapter has considered a wide range of topics which relate to the analysis of tourism within a global framework. None of the topics could be examined with the depth it deserves and all merit further investigation, both as issues pertinent to economics and tourism in general and in the context of specific tourist origins and destinations. The discussion has shown that international tourism can be examined using economic theories relating to trade, industrial economics and growth. Building on the analysis of tourism supply in Chapters 4 and 5, this chapter has shown how market structure is important in determining patterns of specialization in production and trade at the international level. Within the competitive market structures which apply to some components of tourism as well as to other products, relative technological efficiency and/or factor endowments are key variables. The earlier discussion of the determinants of trade indicated that developing countries can use their natural and cultural resources as the basis for tourism supply. The ready availability of low-wage labour may also contribute to the growth of labour-intensive tourism production. Developed countries also have environmental and cultural resources which attract tourists and are able to organize many tourism activities with less labour-intensive production techniques. Thus, developing and developed countries can use their specific factor endowments to supply tourism products with diverse factor intensities of production.

Under the imperfectly competitive market structures which relate to some components of tourism, product differentiation and market segmentation are of particular relevance. Some general equilibrium models of trade have attempted to take account of trade in both homogeneous products and products of different qualities (Helpman and Krugman, 1993). The pattern of inter-industry trade in homogeneous products can be explained by comparative advantage based on different factor endowments, while intra-industry trade between countries with similar factor endowments is based on product differentiation. Over the longer term, international tourism, like trade in goods, may contribute to changes in market structure so that, for example, an airline which has a national monopoly becomes subject to oligopolistic competition in the global arena. Tourism firms and/or national governments may attempt to alter market structure in their favour by such means as inter-firm integration or strategic policy-making. There has been virtually no application of such ideas to international tourism.

Both the causes and the effects of trade in tourism have been discussed in this chapter in the context of a broad distinction between developed and developing economies, in preference to the distinction between large and small countries and islands which has appeared in some tourism literature (for example, Conlin and Baum, 1995). The size of a country or whether or not it is an island are not important *per se*; rather it is the level of national income and wealth, the associated trading relationships and the proportion of GNP arising from tourism which are of particular relevance to tourism's economic effects. For example, the extent of intra-industry linkages tends to be higher within developed economies and leakages from tourists' and local residents' expenditure tend to be lower, resulting in higher income and employment generation. In contrast, lower-income, developing countries are subject to relatively high leakages from tourist expenditure and the associated income and employment generation are usually lower. Moreover, import leakages may rise over time as local residents experience the demonstration effect of tourist spending.

The short- and long-run effects of tourism may be examined using different theoretical approaches. Research on competition between tourism firms at the international level can use theories from industrial and international economics. Studies of the effects of tourism in changing the structure of employment illustrate the insights which feminist as well as a mainstream labour economics perspectives can provide. Analysis of the relationship between tourism and economic growth involves conflicting theoretical viewpoints. According to Keynesian theories of growth, tourism results in increased demand which induces higher investment and income. In contrast, neoclassical theories of growth imply that growth rates are not affected directly as tourism is unlikely to play a significant role in increasing labour, capital or technological progress, although it does provide additional foreign currency which may be used to increase the stock of capital. Endogenous growth theories suggest that increases in the levels of education, training and infrastructure for tourism help to prevent the marginal product of capital from falling, thereby contributing to ongoing growth. Higher demand for labour to work in tourism may also result in rising real wages, thereby stimulating investment in more capital-intensive production and maintaining growth. Therefore applied research to investigate the relevance of the different theoretical perspectives is necessary. Studies could also be undertaken on the possible roles of governmental and international organizations in improving education, training and infrastructure, in conjunction with tourism firms in specific tourist destinations. Governments can play a key role in assisting natural as well as human and capital resources to play an optimal role in tourism development. This is particularly important since natural resources are tourism's primary input base and are often open access goods which are either unpriced or are priced at levels which lead to over-use and degradation, thereby threatening the future of the tourism sectors they underpin. These issues will be discussed in the following chapters.

7

TOURISM AND
ENVIRONMENTAL ISSUES

INTRODUCTION

The term 'environmental' economics has entered economics literature only since the early 1970s. It is used generically to embody not only the analysis of the use of exhaustible energy and productive resources (conservation economics) but also amenity use of natural resources (leisure economics – embracing sport, recreation and tourism), as well as in the accepted sense of investigating the economic role of the environment and the associated causes and impact of its degradation through over-use or pollution, or even government policy. Additionally, it is concerned with the instruments for dealing with the impact of these activities, for example, combating pollution. As the natural environment is largely an open access resource, the subject also covers the valuation of both the non-priced and non-use goods and services it provides.

An essential tenet of environmental economics is that the environment can no longer be perceived as separate from other resources. Any human activity affects it and in turn changes in its state have an economic impact. This impact is both spatial and temporal. In the past, much environmental degradation was a local phenomenon but now the sheer scale of many activities means that it is more pervasive in its significance. Moreover, the effect on the welfare of future generations can be considerable so that it is imperative that the time horizon of economic decisions is taken into account. Any environmental issue, in addition to its economic importance, has ethical connotations. Very often, substantial sections of society feel that their quality of life has been impoverished, even if they have not materially suffered. A sense of loss is experienced if, for example, the biodiversity of the planet is reduced. This is felt on behalf of future generations, as well as the present one. Greater awareness of the interdependence of the environment, economic activity and the quality of life raises political, scientific and social issues in addition to those which are more directly economic. Thus, with the passage of time, environmental economics is not only becoming more and more inclusive but also challenging the more conventional economic analysis of demand and supply and the operation of the market based solely on private costs and benefits.

Tourism is almost wholly dependent on the environment if it is defined to include the human-made as well as the natural form. Resources, such as beaches, seas, mountains, lakes and forests, constitute the natural resource base while historic cities, heritage buildings and monuments are the human-made one. These two forms are what might be called the primary tourism resource base and are the essential component of the product. If it were to be degraded in a given destination, it is likely that tourism would decline. Therefore, an evaluation of current environmental issues and developments and their analysis within economics is particularly relevant to tourism.

In this and the following chapter the aim is to introduce environmental economic analysis and show how it applies to tourism. Current debates on environmental issues are examined with particular reference to the translation of principles into tourism policies and their implementation through both regulatory and market-based instruments. After tracing the development of environmental economics and summarizing its principal issues, this chapter outlines the nature of environmental issues in tourism. The content and scope of environmental economics is then examined and its analytical framework considered before identifying and evaluating the impact of economic activity and policies. In Chapter 8, attention moves to the explanation of methods for valuing the environment and an appraisal of the instruments which can be applied to secure environmental protection and sustainability.

THE DEVELOPMENT OF ENVIRONMENTAL ECONOMICS

It was not until the 1950s that environmental questions were seriously raised within economics by Carson (1963), Boulding (1966), Forrester (1971), Meadows *et al.* (1972) and Schumacher (1973). Sustainability has subsequently emerged as an umbrella term, under which virtually all other aspects of environmental economics are subsumed, placing the natural environment and its conservation at centre stage. Originally conservation economics, the foundation of environmental economics, was solely concerned with the optimal use of resources over time, involving the maximization of net returns from them. It focused on exhaustible energy and material resources and considered the role of price, substitution and technical changes as conservation mechanisms. Such a market-based approach was reinforced by evidence (Barnett and Morse, 1963) indicating that the real cost of productive resources had fallen over time, not risen as would be expected if depletion were occurring, although Hall and Hall (1984) question this result in their study of relative prices in the 1970s. Price substitution and technical change were also seen as determinants of the extent to which the use of secondary sources (recycling) would occur. Market economists argued that as long as good husbandry prevailed, few problems would arise with production from renewable resources such as agricultural land and forests. The approach saw

156

market forces as resolving resource problems and has been subjected to both criticism (for example, D.W. Pearce, 1976) and support (Barnett and Morse, 1963; Barnett, 1979). Even the most committed environmental economist accepts that conservation is not the same as preservation and that the basic optimal resource use principle underpins decisions with regard to sustainability.

With the exception of the seminal work of Carson (1963) and Boulding (1966), it was not until the early 1970s, just pre-dating the oil crisis, that conservation economics started to consider the dangers to the natural environment of the continued pursuit of economic growth. In their thesis on the limits of growth, Meadows *et al.* (1972) asserted that population growth, resource depletion, lack of investment to facilitate exploitation of resources, as well as environmental degradation, would undermine market economies by the early twenty-first century. Other writers questioned the conventional renewability wisdom by showing that over-use of resources would impair natural areas' ability to sustain production. For example, H.S. Gordon (1954) and A.D. Scott (1955), who examined ocean fisheries, an open access resource, indicated that over-fishing would cause stocks to fall below a critical level at which they could renew themselves. Others (Kneese *et al.*, 1970; D.W. Pearce, 1976; Daly, 1977) underlined the link between resource use and environmental problems and examined the instruments to alleviate the problems and their effects in a more policy-oriented approach. These studies demonstrated that phenomena such as externalities and the public good nature of the environment, i.e. market failure, underlay basic problems. Accordingly market failure and possible means of mitigating its effects are important issues and are discussed at greater length below.

Somewhat apart from the developments in conservation economics, investigations were undertaken into the impact of using the environment as a sink for waste and its degradation by pollution through emissions from burning fossil fuels and effluent discharges into water courses. Acceptance that the production of goods and services confers benefits as well as generating costs gives rise to the notion of a social optimum, where benefits accruing to society at large equal the costs falling on it, and constituted the foundation for deriving appropriate policy instruments to achieve such a position. However, since many benefits and costs are not priced through the market, techniques for attaching values to them needed to be derived. The development of these techniques has engendered a vast literature in parallel with that arising from the investigation of the non-priced amenity value of the environment. Once values can be assigned to non-priced benefits and costs, it is possible to consider action to enhance the former and reduce the latter and to compare non-priced demand with market demand in making decisions on the allocation of resources. Further research has been undertaken on the instruments which can be applied to achieve this, as will be seen in the following chapter.

It was while many of these growing issues were being investigated, that global problems became of public concern, with increased awareness of the widespread effects of chemical discharges on the planet's environment, for example its climate, forests, ozone layer and water courses, lakes and seas and the detrimental impact of human activity on fragile ecosystems and biodiversity (International Union for Conservation of Nature (IUCN), 1980; World Commission on Environment and Development (WCED), 1987; D.W. Pearce *et al.*, 1989). It was increasingly accepted that unrestricted economic growth, coupled with rapid expansion in population and the degradation of environment, could not continue indefinitely. Sustainable development, essentially representing an updated version of the limits to growth (Meadows *et al.*, 1972), materials balance (Ayres and Kneese, 1989) and small is beautiful (Schumacher, 1973) concepts, emerged as a goal which the world's economies should strive to achieve. Thus, from narrowly defined market-based conservation economics, the boundaries of environmental economics have been extended both to a more general level of concern and to very specific levels. It is against this background of the development of the subject that environmental issues in tourism are examined.

ENVIRONMENTAL ISSUES IN TOURISM

International business and holiday tourism movements in the early 1990s have grown at an annual rate of around 5 per cent and total over 500 million (World Tourism Organization (WTO), 1992). Domestic tourism is estimated to be ten times this number. Major economic, environmental and social impacts follow as increasing numbers visit not only established but also lesser known and more remote destinations, drawn by the attractive environment, recreational and sporting resources and culture. In addition to these primary resources, tourists require secondary supporting ones, such as accommodation, transport facilities, shops, restaurants and other services which entail physical changes in the destination, including expansion of the built environment.

Like any other productive activity, tourism consumes resources. Given that it is one of the major economic activities in the world, contributing about 6 per cent of global income in 1993 as estimated by the World Travel and Tourism Council (WTTC: see Lundberg *et al.*, 1995), it has a marked impact on the demand for exhaustible and renewable resources. It generates significant wastes which, although not as hazardous as much pollution from heavy industry, manufacturing and chemicals production, can create acute disposal problems (Stabler and Goodall, 1996) as well as major environmental problems. The operation of tourism firms reflects the market-driven characteristics of other sectors where the environment is treated as a free good, in this case an essential input as well as an element of the final product, so that problems occur, especially over-exploitation of the natural resource base and the generation of non-priced adverse effects. Much tourism expansion, particularly its concentration in certain areas, has neglected the long-term dependence of the industry upon the environment (Cater and Goodall, 1992).

The environmental effects, if widely defined so as to include cultural and social elements, are probably the biggest problem of tourism. Areas where overcrowding and overdevelopment occur are often relatively small and possess fragile environments. At peak times, visitors can outnumber the resident population by a factor as high as three or more. Hosts, tourism firms and tourists are seldom aware of the damage being caused; indeed it is usually unintentional, for example, compaction of snow on ski runs which damages plants and the ecosystem (Tyler, 1989) or the fading and deterioration of paintings and murals in galleries and heritage buildings because of greater exposure to strong light and body moisture (Goodall, 1992). Other effects are more deliberate, for instance off-road vehicular use (Sindiyo and Pertet, 1984). Excess numbers also increase the demand for secondary resources, water and energy which may be scarce in developing countries and islands (Romeril, 1989). Loss of flora and fauna occurs where tourism expansion (Andronikou, 1987), climbing (Pawson et al., 1984) and hunting (C. Smith and Jenner, 1989) have taken place. The influx of tourists with a different lifestyle, large financial resources and demand for non-indigenous services not only disturbs existing economic systems, but also can destroy traditional cultures (D.G. Pearce, 1989).

These problems have been recognized by many involved in tourism and have become issues of concern (Cater and Goodall, 1992; Eber, 1992; Jenner and Smith, 1992). The attainment of sustainable tourism is seen as one of the most urgent issues but its meaning has often been misunderstood or distorted to accord with commercial aims. It is also used as an all-embracing term to include what should be separately identified as issues requiring specific solutions, for example resources conservation, waste disposal management and pollution control. Considerable attention has been paid to what is called 'eco' or 'green' tourism (see, for example, Cater, 1993). There has also been increasing concern over the impact of tourist demand on wildlife. Calls have been made for tourists to be more responsible in their behaviour while on holiday (Krippendorf, 1987) both with regard to the environment and their hosts' lifestyle and culture.

Although the environmental impact of tourism is most visible in destinations, there are also effects in origins and while tourists are in transit. For example, the output of aircraft, ferries, coaches, cars, equipment and promotional material consumes productive and energy resources and generates waste in origin areas while travel creates pollution in the atmosphere and adversely affects the environment of areas traversed. The difficulties of taking co-ordinated and concerted action, at least to mitigate if not eliminate environmental problems, are formidable. It is also apparent from inspection of the literature on tourism that most discussion in the mid-1990s has been concerned with identifying the causes and effects of problems. In practice, combating problems has largely been confined to limited initiatives by certain sectors of tourism, notably the hospitality sector (Dingle, 1995; International Hotels Environment Initiative (IHEI), 1993). It is not unduly critical to assert that most utterances by other sectors pay only lip-service to developing environmentally responsible behaviour (Brierton, 1991; Hunter and Green, 1995). This is, in part, a reflection of the failure by governments to commit themselves fully to pursue environmental policies, despite being signatories to international agreements. However, within tourism, as in other industrial fields, as well as a lack of comprehension by businesses of environmental issues and objectives (Stabler and Goodall, 1997), an element of complacency is engendered in that in some areas of activity, it is known that the potential to attain sustainability is limited. Furthermore, firms lack knowledge of the implications, in monetary terms, of investment in, operation, monitoring and management of systems to achieve specified objectives (Institute of Business Ethics, 1994; Forsyth, 1995). Understandably, the private sector tends to perceive the introduction of environmental practices as adding to costs.

Economics can play a central role at a strategic level by informing practitioners of the issues and the consequences of current economic trends and their environmental impact. At a more applied level, the subject can indicate, particularly through its analytical framework, methods of valuing benefits and costs and market-based instruments for attaining environmental objectives, including quality (Stabler, 1996a; Keane, 1997). It is in a position to provide the necessary data and analysis for business people to make informed decisions, although the potential has not yet been realized. Economics can also make a contribution to the largely applied tourism literature, which comprises few texts containing rigorous economic explanations of environmental problems and issues. The dearth of texts which incorporate developments in environmental analysis occurring in mainstream economics is, in part, because of the limited interest of economists in tourism. Economic approaches have focused on specific areas largely found in research reports and journal articles or they have contributed general introductions to tourism economics, such as Bull (1991), Tribe (1995) and Lundberg *et al.* (1995), which do not consider environmental issues in detail. Another reason is that environmental concerns in tourism have largely been examined by non-economists who have seldom been made aware of economic theory and its applications (Burns and Holden, 1995).

One aspect of tourism which economists have analysed and which has implications for environmental change and damage is economic development and its impact. The economics literature has concentrated on estimating income and employment generation and foreign currency earnings which developing countries gain from international tourism and the implications for their balance of payments (for example, Archer, 1977b, 1989; Baretje, 1988; G. Lee, 1987; D.G. Pearce, 1989). The studies have not considered the impact of environmental deterioration on demand and thus income, employment and currency receipts, or estimated the social costs of tourism's role in economic development, either environmental or other. Equally, there have been few studies directly related to the possible beneficial effects of tourism on the environment in some areas; for example, conservation of fragile ecological areas and the designation and, more importantly, maintenance of wildlife reserves. In this context, it is clear that some conservation areas, such as African game parks, are sustained because they are seen as valuable tourism resources. Most economic studies of relevance have arisen from the amenity demand for or conservation of natural, mainly rural, environments and the need to attach a value to them in making resource allocation decisions. The other areas of economic analysis are the techniques for measuring costs and benefits and the instruments for effecting environmental improvement and their evaluation. They, too, have been developed in economics with virtually no reference to tourism and will be considered in Chapter 8.

THE CONTENT AND SCOPE OF ENVIRONMENTAL ECONOMICS

It is against the background of the development of environmental economics and its application to tourism that the issues contained in the rectangles in Figure 7.1 have been identified. In the figure an attempt has been made to reflect how increasing awareness of environmental problems has expanded the scope of environmental economics and to indicate the interrelationships between what hitherto tended to be seen as separate areas of study. No claim is made that the figure is comprehensive. Indeed some areas of environmental economics, for example, the analysis of such concepts and issues as the precautionary principle, the uncertainty of outcomes of particular actions or the effect of policy initiatives cannot easily be depicted in it. Nor is the figure meant to show the interrelationship between consumption, production and the environment as presented in economics texts (for example, Turner *et al.*, 1994); it is more concerned with suggesting areas of its study within economics. Nevertheless, by implication it indicates that some aspects are embodied within others in the sense that, in general, a wider perspective is indicated at the top of the figure with more focused and detailed research interests at the bottom. It is also unavoidable that some elements of the subject spill over into consideration of substantive problems rather than conceptual and methodological ones.

Figure 7.1 is underpinned by the fundamental economic presumption that individuals act rationally (i.e. consistently) to maximize their self-interest. The value of goods and services is expressed through prices. Choices have to be made because resources are scarce in relation to unlimited wants. However,

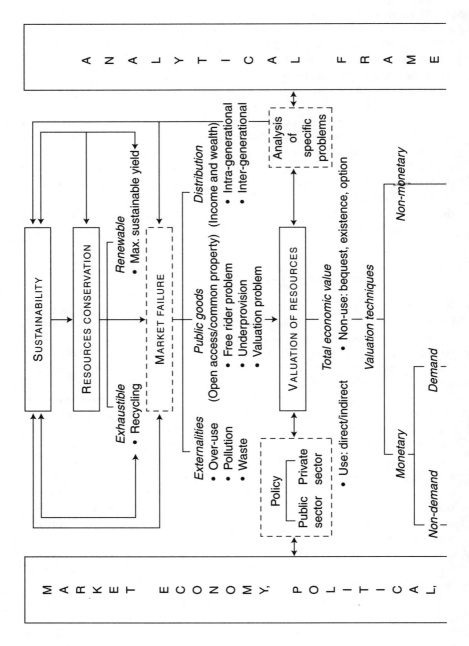

ANALYTICAL FRAME

MARKET ECONOMY, POLITICAL,

SUSTAINABILITY

RESOURCES CONSERVATION

Exhaustible
• Recycling

Renewable
• Max. sustainable yield

MARKET FAILURE

Externalities
• Over-use
• Pollution
• Waste

Public goods
(Open access/common property)
• Free rider problem
• Underprovision
• Valuation problem

Distribution
(Income and wealth)
• Intra-generational
• Inter-generational

Analysis
of
specific
problems

VALUATION OF RESOURCES

Total economic value
• Non-use: bequest, existence, option
• Use: direct/indirect

Policy
Public   Private
sector   sector

Valuation techniques

Monetary

Demand

Non-demand

Non-monetary

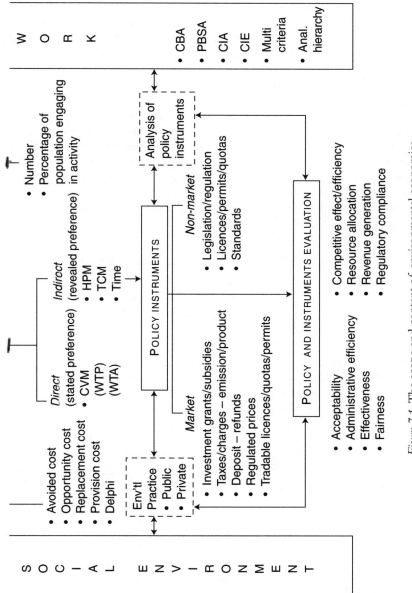

Figure 7.1 The scope and content of environmental economics
Note: An explanation of the abbreviations is given in Chapter 8.

environmental economics posits that this conventional market analysis is at odds with the long-term maximization of welfare because it is based on restrictive assumptions and ignores some phenomena relevant to sustaining human existence. In effect, it is argued that economic analysis should encompass preferences for goods and services which are not traded in markets, for example, the amenity demand met by the natural environment. Thus the context in which environmental economics is set is shown by the vertical rectangles in Figure 7.1. On the left the market economy, political and social environment is shown as a dominant feature determining human activity and behaviour. On the right, it is suggested that all actions affecting the environment need to be appraised within some kind of analytical framework, the justification for which emerges in the examination of each of the major issues. The horizontal rectangles in Figure 7.1 relate to key areas of environmental economics and below these, relevant aspects coming within the purview of each area are identified, by a square or bullet point in some cases where examples are included. The selection of the key areas reflects the development of the subject as well as illustrating its concepts, methods and analytical techniques which are of significance for tourism.

THE OVERALL ANALYTICAL FRAMEWORK

The ensuing sections of the chapter consider, first, the analytical framework which economists have advocated for the consideration of environmental issues and action on them. Then the broad conceptual issues of sustainability, resources conservation and market failure and its consequences are examined. The remaining elements of environmental economics shown in Figure 7.1 are discussed in the following chapter.

The essence of the economic analysis of resources use is that their cost for a specific activity should be expressed in terms of the opportunity cost, i.e. the benefits forgone by not using them for an alternative activity. An associated guiding principle is that the benefits should outweigh the costs, otherwise there is no point in continuing to employ the resources for a particular activity. In the private sector an appraisal of a project would include only the direct monetary returns (benefits) and costs to establish whether an adequate rate of return can be secured. Other than for minor short-term investment, a discounted cash flow would form part of the appraisal. The internal rate of return on capital employed, over the life of the investment, can act as a measure which should at least equal the external rate of return, i.e. the market interest rate which represents the opportunity cost of funds invested within the enterprise or project.

Given the nature of environmental resources and problems, for example, the value to society of the conservation of scenic areas or heritage buildings which could alternatively be used for purely commercial activities, it is inappropriate to use appraisal methods which take account only of costs and benefits which are priced in the market. Economists have derived cost-benefit analysis (CBA) as a suitable framework for the assessment of projects which involve both monetary and non-monetary costs and benefits, as well as instances of large capital outlays, a very long time period over which costs and benefits accrue and a discount rate which reflects social time preferences. Discounting carries the implicit assumption that present consumption is preferred to that in the future, raising the question of whether a 'perfect conservationist' would advocate a zero discount rate.

The CBA method originated in the 1930s in the United States where there was a need to show not only the direct value of public expenditure on flood prevention programmes to agriculture but also the wider indirect benefits to others likely to be affected by floodwater. The method has developed since then to cover education, health and transport programmes, energy use, mineral extraction, pollution, environmental quality and recreation (see, for example, standard texts by Marglin, 1967; Layard, 1972; I.M.D. Little and Mirlees, 1974; D.W. Pearce and Nash, 1981; Hanley and Spash, 1993). Despite its increasing sophistication, there are still difficulties in applying CBA in practice. It does not fully cover all cases and it has critics who perceive fundamental flaws in a number of techniques employed within it (Mishan, 1971). Many problems encountered with CBA largely stem from forcing specific studies to conform to the standard method rather than clearly defining and delineating the project and then adapting or deriving a version of CBA within which to conduct the appraisal. Urban conservation is a case in point because of the characteristics of the built environment in terms of its function, physical structure, location, surrounding environment and legal status, as well as the many cultural, ecological, economic, environmental and social effects of an intangible nature which are generated.

Lack of space precludes a full exposition and review of CBA and its development here but it is useful to note refinements of the basic method and adaptations to meet specific circumstances. Cost-benefit analysis and related methods have been refined in order to provide cheaper, more useful alternatives in practical contexts. The closest to CBA is the planning balance sheet analysis (PBSA). Lichfield (1988) further developed this approach, originally put forward in the mid-1950s, into a Community Impact Analysis (CIA) or Community Impact Evaluation (CIE). The PBSA was explicitly devised to overcome the fact that many costs and benefits are not easily measured in money terms, so that the results of any social cost-benefit analysis are always liable to objections that some costs or benefits were incorrectly valued. Thus, the PBSA approach stops short of assigning values to many costs or benefits, simply indicating where they should be placed on the balance sheet, either as assets or liabilities. The CIA approach further indicates which sections of the community are likely to gain or lose from a decision, so taking into account the distributional effects as well as what might be called the efficiency effects.

Another type of approach is the multi-criteria analysis developed by Nijkamp and others (Nijkamp, 1975, 1988; Paelinck, 1976; Voogd, 1988).

Here, the various options are ranked according to criteria thought to be relevant and the 'best' chosen through calculating the extent to which it outranks others on average. A more mathematical approach to decision making between alternatives is the analytic hierarchy process developed by Saaty (1987; see Zahedi, 1986), which has been suggested for the evaluation of alternative approaches to conservation and restoration by Lombardi and Sirchia (1990) and Roscelli and Zorzi (1990). This, too, is designed to formalize the process of choosing between alternatives in the absence of full information. Increasingly, risk analysis is being incorporated into such evaluatory and decision techniques.

In environmental economics an overall analytical framework is required which

1 allows for the assessment of all significant benefits and costs; that is those which

- are priced
- can be evaluated in monetary terms even if not normally traded in the market
- cannot be evaluated in monetary terms but can be measured quantitatively
- cannot be measured quantitatively

2 takes account of both inter- and intra-generational distribution effects
3 avoids double-counting
4 takes account of the impact on the allocation of resources, especially where these are unique, irreproducible and where any decisions may be irreversible
5 clearly identifies and evaluates the impacts of actions
6 considers the implications of the objectives of the appraisal and for those conducting it
7 takes cognizance of the policies which influence or are employed to achieve the aims of the appraisal.

Such a framework, particularly criteria 4, 5, 6 and 7, widens the scope of conventional CBA in the global, very long term and sustainability context (Stabler, 1995a).

CURRENT ENVIRONMENTAL ISSUES IN ECONOMICS AND TOURISM

Sustainability

Sustainability is generally perceived as a term embracing all aspects of economic and environmental matters, having acquired a wider meaning than the term 'sustainable development' (SD) which is normally used. The definition

166

of SD frequently quoted is that agreed by the Brundtland Commission (WCED, 1987) but D.W. Pearce *et al.* (1989) allege that over 300 alternative definitions have been suggested so that it is not surprising that much confusion has occurred, particularly among practitioners. As a result of what has become known as the Rio Declaration in 1992, SD has become more a set of principles for actions and agreements on the environment in which biodiversity, climate change and forest management and conservation were accorded prominence, alongside the priority to be given to the poorest section of the world's population. The essence of SD is to manage world economies to attain intra- and inter-generational equity. This means that present needs should be met without impairing the capacity to meet future needs. However, there are two major and differing interpretations as to what this signifies. The weak approach to SD accepts that as long as the total stock of capital, i.e. the natural environment plus human-made resources, is not depleted in total, then substitution of the latter for the former is permissible. This implies that the natural environment might be degraded. The strong approach opposes this by insisting that no substitution should be allowed. Indeed, the extreme strong approach goes further by asserting that degradation is already unacceptable and that steps should be taken to improve the quality of the environment.

Whatever stance is taken between the weak and strong positions, the implication of SD is that economic growth rates should be moderated. Moreover, it is posited that the variables used at present to measure welfare and changes in it, most often income per capita, are unsatisfactory. The quality of life, which cannot always be evaluated in monetary terms, should be taken into account. Reductions in adverse externalities such as chemical pollution or noise levels, increases in life expectancy and an enhanced physical environment are examples of quality of life variables. In the face of uncertainty, particularly with regard to the ability of the environment to assimilate pollution, or the possibility of taking actions which are irreversible, it is argued that the precautionary principle should prevail. This principle has not been unequivocally defined but in essence, within the context of sustainability, it posits that given uncertainty and possible irreversibilities, to conform to the intergenerational tenet of constant natural capital and capital bequests, no action should be taken which results in capital depreciation. Other prescriptive principles are that the full costs of production, i.e. those which take account of the social cost generated, including environmental costs, should be taken as the appropriate measure of the cost of resource use. As far as possible renewable resources, consistent with their ability to regenerate themselves, should be substituted for depletable resources. Paradoxically, it is salutary to note that over recorded history, the only resources which have been totally extinguished are those which ostensibly should be renewable.

A fundamental issue associated with SD is how to reconcile economic development and growth with the open access public good nature of the natural environment which consequently suffers from detrimental externalities. In a sense, SD is a very specific economic concept. It is possible to conceive of various types of sustainability, for instance sustainable agriculture, sustainable cities, sustainable ecological systems and sustainable tourism. These are sectoral variants of sustainability, which is the ability of systems to perpetuate themselves in the very long run. This does not necessarily suggest a steady-state because change is inevitable, even if humans cease to exist.

Sustainable tourism

Sustainability should be the cornerstone of the development of the tourism industry since the natural environment constitutes most of its primary resource base. Moreover, with growing anxiety over environmental deterioration on the part of both tourists and residents, firms and governments are under increasing pressure not only to endorse sustainability principles but to encourage positive action to bring it about. Reflecting the advocacy of sustainable development (SD) in academic and practitioner circles, the term sustainable tourism (ST) has been coined. However, it is clear that neither the SD nor ST concepts have been fully understood in most of the business world. There is considerable evidence that in developed countries tourism firms, in common with those in other commercial sectors, subscribe largely unconsciously to the weak interpretation of SD (Cook *et al.*, 1992; Global Environmental Management Initiative, 1992; Hemming, 1993; Welford and Gouldson, 1993; Beioley, 1995). This implies a continuing deterioration of the natural resource base. Tourism practitioners' understanding of ST is 'viable tourism' in the commercial sense that businesses are profitable and will survive (Stabler and Goodall, 1992, 1996; Wight, 1993, 1994; Stabler, 1995b; Forsyth, 1995). This attitude and approach to ST tends to be reinforced by central and local governments and tourism bodies, which often view tourism as a vehicle for diversifying and developing their economies and achieving higher revenue.

The actions taken by firms (considered on pp. 174–7) are concerned with conservation of energy and materials resources and the minimization of waste as a means of cutting costs or increasing revenue and profits. Being seen to be 'green' enhances a business's image and may present market opportunities for increasing demand. Indeed, firms have taken the concern by consumers that tourism should be environmentally responsible and turned it to commercial advantage by promoting eco and green holidays. Environmental actions which are not motivated in this way are likely to come about only if firms need to comply with environmentally related regulations. There is, as yet, no coherent strategy on sustainability because past incentives have mainly been geared to tourism expansion. Moreover, the largely fragmented structure of tourism, whereby decisions and investment emanate from numerous independent firms, militates against co-ordinated policies and practices. The issue is thus a complex one which only the public sector has the potential for resolving comprehensively on a 'top-down' basis.

Economics makes clear what SD and therefore ST entail in terms of action. It demonstrates the case for intervention in the market and shows why. The concept of market failure, the principles of resource conservation, the notion of a social optimum, the valuation of social benefits and costs and the analysis of price-based instruments for achieving environmental improvements are all elements in attaining the goal of sustainable tourism, as will be revealed in the examination of these issues.

Resources conservation

The rate of depletion and possible exhaustion of key productive resources remains a central economic problem. The literature on conservation is also concerned with the consequences for growth, the possible solutions, for example technological developments, and the role of market costs and prices in inducing conservation through reduced demand, substitution and more efficient use of resources, including the extension of product life and recovery. Since the late 1970s, however, the emphasis has shifted somewhat to assessing, for exhaustible resources, the opportunity cost of exploiting primary as opposed to secondary (recovered) and waste (recycled) materials sources. Recycling, which has seized the imagination of consumers, is viewed rather more sceptically by economists, mainly because of the costs involved in the recycling process which are not always fully identified and may outweigh the benefits. With respect to renewable resources, a dominant issue is the implications of the open access characteristic of much of the natural environment for the survival of many species of flora and fauna, the sustainability of some sources of food and materials and the conservation of amenity resources. Maximum sustainable yield, an essentially biological concept and a long-standing basis for the modelling of renewable resources in economics, has surprisingly not figured strongly in the more strategically oriented focus of SD, nor in ST. Yet this concept has strong practical significance in implementing SD and ST principles and policy. Recycling and the notion of maximum sustainable yield are, thus, identified as important issues in their own right, as well as having implications for tourism as a major user of resources and with a vested interest in the conservation of renewable resources.

Recycling

A distinction can be made between the terms 'recovery' and 'recycling'. Normally, in the context of recycling, recovery rates mean the proportion of a material from a primary source which can be made available for re-use. The extent to which recycling takes place is influenced by the nature of the material, at what stage in the product cycle recovery occurs, who uses it, what proportion is unusable residual waste and how it is disposed of and the ease or difficulty with which it can be recovered, one crucial determinant being the scale of the operation. These physical factors determine the costs of recycling, which may be considerable, but an important additional variable is the price of the material which is governed by demand in relation to supply.

In general, economists argue that recycling becomes feasible if the costs of recovery are lower than those of exploring for and exploiting primary sources. However, market prices and costs do not reflect the true benefits and costs: while they take account of the private costs and benefits they do not embody the additional non-priced costs and benefits, i.e. externalities. For example, the recovery of materials from discarded cars involves their separation and those which have no value have to be disposed of by combustion or landfill, creating detrimental externalities, such as the production of harmful emissions, toxic chemicals and sterilization of amenity and developable land. On the other hand, if a higher recovery rate is attained, which in the example of cars might depend on incorporating a larger percentage of recyclable material when they are first manufactured, unusable waste is reduced. Consequently, the social costs generated by primary resource utilization are lessened, such as the despoliation of scenic areas where metal ores are mined, the generation of noise and dust and associated health hazards, because reprocessing recovered materials normally creates fewer externalities. Moreover, there is the concomitant benefit of a reduction in the disposal of waste with its accompanying environmental impacts.

The problem of waste generation in relation to the primary and secondary resources use also raises the issue of the full costs of its disposal. Ignoring its externalities effectively means waste disposal is underpriced. In addition, in the interest of public health, central and local government intervene to set standards or regulate or levy charges, involving administrative costs. Intervention distorts prices and costs while administrative expenses are often not fully assigned to the service, so that the full costs are not passed on in user charges. The inference is that more recycling would occur if the full costs of disposal had to be met by users of such services. These arguments are relevant to tourism in that it is a substantial consumer of resources and generator of wastes. Furthermore, many destinations are in islands or small states, often set in fragile environments. Reductions in the use of primary sources of materials combined with recycling can make a significant contribution to achieving sustainable tourism (Stabler and Goodall, 1996). Tourism associations recognize this contribution and urge businesses to institute recycling programmes (Troyer, 1992; IHEI, 1993; WTTC, 1994).

Economics, by identifying and evaluating the full social costs and benefits, i.e. the private costs and benefits and externalities of resource use, including the incidence of recovery, is able to indicate the optimal level of both primary exploitation and recycling. The concepts of private, social and ecological optima may be illustrated diagramatically, as shown in Figure 7.2.

It is assumed that a tourism activity generates both marginal social benefits (MSB) and marginal social costs (MSC), which exceed private benefits (marginal revenue: MR) and marginal private costs (MPC). The private optimum is at Q_P where MR = MPC whereas the social optimum is at Q_S, suggesting that a reduction in the level of tourism activity from Q_P to Q_S is necessary. However, this social optimum does not equate to what can be called the ecological optimum indicated by Q_E, beyond which the environmental assimilative capacity (EAC), shown by the horizontal broken line, is exceeded, generating an unassimilated detrimental externality abc. Should this be cumulative, then environmental degradation increases in each succeeding period. The conceptual distinctions between private, social and ecological optima are easily identified but deciding which should be pursued is crucial in terms of minimizing tourism's wider detrimental effects and the attainment of long-term sustainability. This issue is examined in Chapter 8.

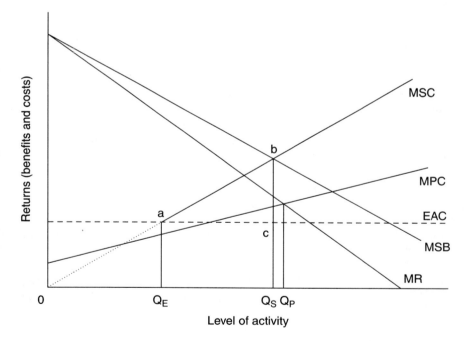

Figure 7.2 The concept of an economic optimum as applied to tourism
Note: Q_E = ecological optimum, Q_S = social optimum, Q_P = private optimum, MSC = marginal social cost, MPC = marginal private cost, MSB = marginal social benefit, MR = marginal revenue, EAC = environmental assimilative capacity

Maximum sustainable yield

The concept of maximum sustainable yield is concerned with resources which are capable of renewal either naturally or through appropriate management. A key issue is how to achieve maximum yield but maintain sustainability from the economic use of open access and common property resources. The former are resources such as the atmosphere and oceans which no one owns, while the latter are resources which are group-owned and managed, often by a local community. Both are susceptible to over-exploitation, although the problems are more acute with open access resources, particularly the dangers of the extinction of wildlife species or degradation of ecosystems. There are many examples of over-use of oceans through excessive fishing which has reduced the stocks of once common species to levels which cannot be sustained. Other more emotive examples are seal and whale hunting. All are relevant to natural resources-based tourism, such as game safaris, seal and whale viewing, snorkelling and scuba-diving.

The economics of maximum sustainable yield considers the relationship between the price of the product, the cost of exploiting it, the yield in terms of the physical quantity and the stock or population. The yield is determined by the exploitation effort and the total stock of the resource. The greater the effort, for example increasing the number of hours worked or the efficiency of the equipment, the greater the yield. However, the size of the resource stock governs its rate of renewal or recruitment and thus the quantity available. The problem becomes an economic issue by considering the revenue generated in relation to the cost of the effort. Normal profit-maximizing conditions can be applied which will determine the quantity of the total stock of the resource taken, which may be below or above that necessary for the resource to renew itself. If the cost of exploitation is reduced or the price of the product rises, or both, then profits will increase which will, in turn, attract more entrants. Since access is open, i.e. free apart from the costs of exploitation, it is likely that exploitation is beyond the point of maximum yield. Should the total stock fall below a given threshold (a biological or ecological issue) then the population will crash, leading, in the case of animal or plant species, to extinction.

The elements of maximum sustainable yield, explained in general terms above, are applied to tourism in Figure 7.3. It is also applicable to land-based open access resources and common property resources should the owners not adopt, or abandon, sustainable management practices. In the conventional exposition of renewable resources, such as fishing and forestry, the shape of the total revenue curve is dictated by the underlying biological relationship between the resource stock, its rate of reproduction or renewal and the harvest rate which is determined by the harvesting effort (see Norton, 1984, or Conrad, 1995, for a full explanation). It is posited that there is an

172

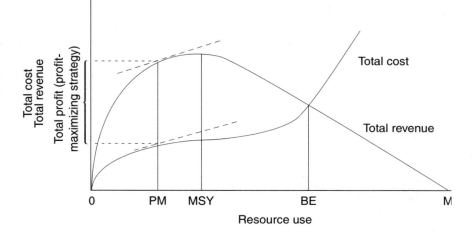

Figure 7.3 A bioeconomic model of maximum sustainable yield applied to tourism use

inverse relationship between the stock and effort. The total cost curve corresponds to that usually constructed in economics, in which increasing returns prevail over a range of output, so flattening the curve before decreasing returns set in, giving rise to a steeper curve. It is also assumed that normal profit, i.e. that required to keep firms in business, is included in the cost curve.

Bioeconomic models can be reinterpreted within the context of tourism. Suppose tour operators and tourism firms in destinations make use of natural resources which can be degraded by over-use. In Figure 7.3 it is assumed that profits are maximized at PM where there is the greatest vertical distance between the total revenue curve and the total cost curve. The resource, however, is capable of sustaining a larger number of users or visitors and generating greater total revenue but at a declining rate of profit to the right of PM, given the shape of the cost curve. The point MSY indicates the maximum revenue possible which declines to zero at point M because of a reduction in the resource's capabilities to support visits. This could occur because of visitors' perceptions of a decline in its quality or as a result of deterioration in its physical properties, for example, a ski resort where the ability of the snow cover to support the activity declines with increasing use. Another possible instance is the case of safari tourism where increasing tourist numbers disturb the breeding patterns to such an extent that the fauna reproduction rate declines.

The level of BE shows the break-even point for providers meeting tourists' demand. This presupposes that the resource is a free access one so that apart from the costs of marketing holidays (which the shape of the cost curve suggests is subject to eventually decreasing returns) there is freedom of entry. Thus, suppliers will continue to exploit the resource even though abnormal profits are competed away, for example, guides in wildlife habitats where over-supply diminishes the earnings of individuals. Should the marketing costs be lower, then the nearer to point M the use of the resource will be, threatening its renewability. On the other hand, constraints on continued exploitation of the resource base sometimes occur through limits on accommodation and facilities. Overall, the reinterpretation of bioeconomic models with open access shows that over-exploitation can occur. Therefore, the economic analysis of maximum sustainable yield holds implications for the management of open access resources, particularly the need to secure workable agreements on resources of global significance. For example, attempts have been made with respect to the oceans regarding fishing and whaling, not always successfully in contrast to the international accord on Antarctica which appears to be working. The issues of classifying natural resources and the problems of their management have been explored by a number of writers and are reviewed by Berkes (1989) and Stabler (1996b).

From these explanations of recycling and maximum sustainable yield, in which reference was made to externalities and open access resources, some vertical links between issues in Figure 7.1 have been traced. Recycling, by slowing down the rate of depletion of exhaustible resources, and the concept of maximum sustainable yield, by indicating how conservation of renewable resources might be achieved, support the goal of sustainable development and sustainable tourism whereas the introduction of externalities and open access resources leads downwards to consideration of market failure.

RESOURCES CONSERVATION AND TOURISM

Exhaustible resources

In the foregoing discussion of sustainability and tourism firms' response to the debates about it, the nature of firms' initiatives was not made explicit, except to suggest that they tended to be driven by market considerations as to the additional costs, or conversely cost savings, and possible generation of revenue and increased profits. Inspection of the literature (for example, International Hotels Environment Initiative, 1993; Middleton and Hawkins, 1993; Scottish Tourist Board, 1993; Beioley, 1995; Dingle, 1995) reveals that environmental actions by firms, particularly in the hospitality and transport sectors, resemble the actions of households. These are to avoid the use of materials likely to be environmentally harmful, substitute purchases of recycled products for those from primary sources and to reduce waste by an absolute cut in the consumption of materials and energy and by recycling materials as far as is practicable. Rather more ambitious are decisions to use only suppliers who meet the environmental standards expected by purchasers and action to educate and encourage tourists to adopt environmentally responsible behaviour.

This 'bottom-up' approach to conservation has not been entirely thought through so that the total benefits and costs have not been taken into account. Firms have seized on the conservation element of the sustainability issue because of the expectation of higher revenues, such as from demand by eco-tourists or financial benefit, for example, the cost savings from better energy and water management and waste minimization effected in hotels (D'Amore, 1992; Hill *et al.*, 1994). If cost savings are passed on in lower prices and consequently attract more tourists then, in total, energy and materials use and waste generation may be no lower in a particular establishment or destination. Moreover, other forms of environmental damage may occur, such as over-use of natural resources and pressure on local services in the area in which a hotel is located. These externalities remain outside the responsibility of individual businesses. Thus, while moves by tourism firms towards the conservation of materials and energy resources reflect the market-oriented approach, conservation has not been fully considered in the wider context of sustainability.

Renewable resources

The emphasis on the issue of maximum sustainable yield and its implications for renewable resource use in economics is of great significance in tourism, given its reliance on the natural environment. Some examples of environmental degradation arising from tourism, in particular its use of open access resources, have already been cited but it is necessary to acknowledge the differences between its human-made and natural resource base and the difficulties of distinguishing exhaustible and renewable resources. Except where they are easily reproduced, many human-made resources, especially heritage artefacts, are unique and so constitute exhaustible resources. Some natural resources, such as the Giant's Causeway, Grand Canyon and the Himalayas, are irreproducible and so are similarly categorized. It is essentially the flora and fauna which are renewable resources but also potentially exhaustible if improperly managed. There are many such resources of this nature which are vital to tourism, for example national parks and game reserves, nature reserves and forests, wetlands and uplands. In addition, there are numerous and varied landscapes, sometimes influenced by human activity which, although not actively utilized, form a backdrop to tourism and are a key component of the product for destinations. Both tourism firms and governments often fail to consider the total nature of the product offered and the impact of tourism on the natural resources component, prioritizing the short-term benefits which they obtain from tourism. Businesses and residents of destination areas, although accepting the demands made on them in order to enjoy the benefits arising from tourism, may be unwilling or unable to become involved in decision-making concerning the environment. Accordingly, the conservation of their environment is compromised. Those who do become actively engaged in protecting their environment are more likely to oppose the expansion of tourism *per se*. These issues have been discussed in the literature under the 'community approach' to tourism development as espoused by, for example, Murphy (1985a) and Keogh (1990) and is paralleled in the natural resources management field by Berkes (1989).

An interesting example of tourism-related wildlife management supported by the local community is the CAMPFIRE programme (Communal Area Management Programme for Indigenous Resources) in Zimbabwe (Barbier, 1992). This involves the establishment of hunting quotas for wildlife and the allocation of licences to safari operators who organize the activity. The revenue from the hunting and sales of meat, hide and skins is used to finance recurrent expenditure including salaries and vehicle costs. A considerable proportion is also used for expenditure on such projects as clinics, crèches, footbridges and housing and the local community receives additional benefits in the form of cash and compensation for animal damage to people and crops. Meat sales are undertaken at subsidized prices, raising nutritional input. Although in its infancy, CAMPFIRE has attracted widespread international interest as a community-based programme which has the potential to be self-sustaining over the long run.

The so-called bioeconomic principle of maximum sustainable yield, which is based on the usual economic optimizing conditions but recognizes that biological or ecological processes govern the availability of the resource, is relevant to ST on two counts. First, it defines the concept of sustainable development/tourism and renders it more comprehensible to the practitioner. Second, it is a significant and attainable step towards sustainability, moving environmental action away from the as yet largely ineffective energy and materials conservation and waste management policies on which firms, including some in tourism, have focused their efforts. The notion of sustainable yield and the revenue it generates can be reinterpreted as sustainable capacity in the context of open access environments, both cultural and physical. The prescription economics offers is that the primary tourism resource base, or the parts which comprise open access resources, should be subject to management regimes which ensure that the maximum contribution to tourism and the local economy is achieved while allowing constant regeneration. Economic analysis of the sustainability of open access resources tends to reach the pessimistic conclusion that over-exploitation and degradation will occur in the context of costless access (H.S. Gordon, 1954; A.D. Scott, 1955; Crutchfield and Zellner, 1962; V.L. Smith, 1968; Clark, 1973). Thus, in addition to the need for detailed information of a physical, biological and economic nature to ascertain maximum sustainability, the management of such resources entails control measures which may need to be regulatory, in the case of resources to which access cannot be controlled, rather than the monetary measures normally advocated by economists.

Devices suggested in the literature on fishing (Anderson, 1995; Conrad, 1995) are applicable to tourism resources. Such measures as subsidies to leave the industry or landing taxes can be adapted to tourism whereby subsidies could be paid to transport operators or hoteliers to reduce occupancy while tourists could be taxed for entry to or departure from certain areas. Indeed, charges such as airport taxes and landing fees are levied in some countries. In the Southern Ocean, tourists are charged to set foot in Macquarie Island, the revenue being used to protect the environment. This instrument is also used in the Annapurna conservation area. Regulatory devices in fishing include closed seasons, year restrictions, limited entry and catch quotas. The first of these is possible and indeed is already practised in environmentally sensitive areas. The last three devices are easily transcribed into a tourism context where the over-use problem is largely one of numbers. Entry numbers can be controlled as in the case of Venice, for instance by supplying a maximum number of bed-places in a resort or by the issue of a limited number of permits or visas. Bermuda, for example, has exercised control by limiting the number allowed entry each year. Once in a destination, restrictions can be imposed on the demand side, for example, to reduce environmental impacts of skiers, hikers and scuba-divers, and on transport operators (restriction of vehicle numbers cr capacity) and retailers (to stock only certain products on the supply side).

Another aspect of the economics of sustainable yield which is relevant to tourism is the recognition that in the absence of clearly defined property rights, an institutional approach is desirable because it is necessary to negotiate, monitor and enforce agreements. Moreover, significant transaction costs are often involved. These factors have already been considered in Chapter 5 on supply and indicate a need to incorporate institutional considerations within mainstream analysis. In addition, to be successful the management of open access resources entails co-operation. This has led to the application of game theory to such problems, particularly in the face of imperfect information and uncertainty. Issues concerning the effect of high transaction costs, property rights and the role of institutions in the process of attaining sustainable development have hardly been addressed in the environmental economics field so it is not surprising that they have not been considered in relation to tourism, although Hunter and Green (1995) and Stabler (1996c) have touched on them in reviewing the role of planning and resource management regimes respectively. Further reference to renewable and open access resources is made in the following section.

MARKET FAILURE

Market failure, shown within a broken-line rectangle in Figure 7.1, is a core concept in economics, influencing the way that economists view environmental problems and possible solutions to them. The concept effectively underpins all of environmental economics. The inability of markets, where demand and supply are determined by price, to provide some goods, especially environmental ones, either at the level society considers optimal or at all, arises essentially from the public good nature of air, land and water resources. Access to these cannot be excluded unless private property rights can be exercised so that, as a consequence, no price can be charged to those enjoying their benefits. In practice, high levels of consumption tend to reduce the enjoyment of individuals or other adverse effects are generated, although consumption by one person may not reduce that by another. Hence, as shown in Figure 7.1 and demonstrated in considering resource conservation, three features of market failure, externalities, public goods and distribution, are relevant to environmental economics and partly explain the remainder of the figure.

The first feature is the public-good nature of natural and some human-made environments, which gives rise to the problems identified. Public goods are those from which no one can be excluded from consuming or enjoying. Examples often cited in the economics literature are street lighting and amenity open space. The private sector has little or no incentive to supply such goods or services as it is difficult to exclude consumers, for example, access to a mountain or beach. Even if partial exclusion is possible, some individuals known as free riders, if not compelled to do so, would consume them without paying, for instance where payment is voluntary. As a consequence, it is likely that underprovision by the private sector would occur. Enjoyment of public goods also means that willingness to pay and the number of people consuming them free cannot be established as in private markets, thereby creating an estimation of demand and valuation problem. This is an acute difficulty with many natural resources used for recreation and tourism.

Externalities, as a second feature of market failure, are almost an inseparable part of public goods. Any good which is free tends to be undervalued and accordingly abused in so far as the detrimental impact of its use imposes no cost on the individual and so is ignored. Free access can lead to over-use and deterioration, for example, erosion of footpaths, ski runs or fragile ecosystems and natural environments are also used as depositories for chemical emissions and effluents (pollution) and physical residuals (waste) arising from production and consumption, including tourism. A number of examples of the externalities generated by tourists' activities were cited earlier and others are given in Chapter 8.

Less obvious and not always cited as an aspect of market failure is the distributional problem. The Rio Declaration in 1992 identified concern for the poorer nations of the world and the differences in opportunities and lifestyles of the rich and poor. This intra-generational inequality has an environmental dimension with respect to the rate of use of resources, the impact on the natural and human resource base and the attainment of sustainability. It is also an essential element in valuing resources and devising policy instruments. It emerges very strongly as an issue in the appraisal of capital projects within appropriate analytical frameworks. Projects in the private sector usually ignore distributional issues, particularly with regard to environmental impacts. Within CBA and its variants, which are considered appropriate in public sector appraisals, distributional effects can be taken into account by incorporating weighting to reflect relative impacts on different sections of society. For example, it is obvious that the construction of new roads will generate benefits and costs, but on whom will they fall and what proportion of the community will be affected? If the benefits are enjoyed by a small percentage of society, perhaps the rich, while the costs are borne by a higher percentage, maybe the poor, this raises the question as to whether the benefits should carry a lower weighting than the costs. There are numerous examples of intra-generational inequity in tourism, particularly in the discrepancies in the quality of life of visitors and residents of host communities. The organization Tourism Concern has cited exploitation of indigenous communities in developing countries by those from developed nations, for instance, the displacement of lower-income groups from areas where tourist demand has led to soaring property and land prices, as has occurred in Burma, Egypt, The Gambia, Morocco and the Philippines (Tourism Concern, 1995); the destruction of food-growing land and fishing waters in Hawaii (Puhipan, 1994) and the corruption of culture in Fiji (Helu Thaman, 1992).

Without underrating the claims of the present generation, environmental economists, especially where sustainability issues are concerned, argue that consideration of the inter-generational effects of current human activity are paramount. Thus the natural assets rule, equated to the hard approach to sustainability, acknowledges that the present generation holds in trust the natural environment which should be passed on undegraded to future generations. Inter-generational effects reflect market failure as, on the grounds of equity, they are perceived as important but are not taken into account in the operation of the market, simply because the time horizons of the present generation are relatively short so that future benefits and costs are discounted more heavily. They also occur because of the public good and externality problems.

Market failure is not confined to these three features only; monopoly is normally also included. However, in the tourism environmental context it is not as significant, although likely to be increasingly so in the future with the emergence of greater internationalization and concentration in tourism enterprises. The crucial point about externalities, public goods and distributional issues is that their effects constitute justification for market intervention. It is market failure and intervention which are the focus of the remainder of Figure 7.1.

Decisions on the allocation of resources in terms of their opportunity cost require that like should be compared with like. Effectively, such decisions necessitate the attachment of prices to public goods in order to effect comparison with those which are private and are traded in the market.

Externalities, whether detrimental or beneficial, are unpriced elements of market goods. Decisions on distribution indirectly involve the pricing problem because of the difficulty of evaluating, say, the benefits and costs of reallocating resources for use by future generations. The valuation of resources is an essential step in the process of intervening to mitigate the effects of market failure for unless public goods or externalities are assigned values, it is not possible to devise appropriate instruments to effect changes in the market. For example, to institute a tax to reach a social optimum in the use of roads where the marginal social benefits of motoring equal its marginal social costs, it is necessary to estimate the externalities, i.e. the difference between the private and social benefits and costs.

MARKET FAILURE IN THE CONTEXT OF TOURISM

Although it has its critics (A. Randall, 1993), the idea of market failure is firmly entrenched as part of the conventional wisdom of economics and, furthermore, is perceived as the rationale for land-use planning and other forms of government intervention. It is acknowledged that if markets fail to perform efficiently there is a need for ameliorative measures involving governments as law enactors and regulators covering public finance, collective (public) goods and, within environmental economics, the management of natural resources and environmental quality. Within the tourism literature and practice there is some appreciation of externality and public good problems and the issues that arise from them, but there is much less recognition of intra- and inter-generational factors and monopoly. The implications of market failure, in terms of resource allocation decisions, and, thus, the need to measure demand and value non-priced resources, received scant attention until the late 1970s (for example, Hanley and Spash, 1993). Another difficulty is that even in the tourism academic literature, the manifestations of market failure, especially externalities and public goods, are commonly misunderstood and therefore not always explained correctly because they are perceived as identical, being outside the purview of tourism business operations. There is certainly a close relationship between the two; for example, a common access resource which is over-used by tourists can give rise to a detrimental externality, but the two are not synonymous.

Market failure is often introduced as merely signifying inefficiency in the operation of markets and likely not to persist if imperfections and high transaction costs can be eliminated and the creation and definition of property rights accomplished so that the externality and public good problems are resolved (A. Randall, 1993). Accordingly, economists who acknowledge the existence of market failure do not necessarily accept that it is an inherent phenomenon so that some type of intervention is always necessary; in contrast, others believe it to be intrinsic to the operation of markets. The discipline thus sends out conflicting signals to the business world which under-mine attempts to initiate action by either public bodies or those operating in the market.

While some of the problems associated with externalities and public goods have been recognized and understood implicitly by tourism researchers and practitioners, those concerned with intra- and inter-generational issues have not. Recognition that the welfare of neither the present nor future genera-tions is adequately represented in markets is insufficient; the problem goes further than that. There are, with regard to future generations, the problems of imperfect information, the existence of uncertainty and the common human preference for current rather than deferred consumption. The equity issue is an integral part of the arguments for achieving sustainability and, given the lack of clear definitions and understanding of this concept, the weak commitment to it is understandable. What the idea of market failure, as subscribed to by environmental economists, does convey is its potential to highlight basic environmental problems which are likely to remain unre-solved through the unencumbered operation of the market and, therefore, necessitate some form of intervention.

In practice, there have been few attempts to relate the activities of tourists and the operation of tourism firms to policies necessary to deal with envi-ronmental issues which the economic analysis of market failure prescribes. That the full implications of the market failure concept have not been fully grasped by practitioners in general and in tourism in particular is not surpris-ing. The term itself is a source of confusion since the business sector does not perceive that markets actually fail, and it certainly does not fully compre-hend that they do not function in accordance with an economic ideal. This is, in part, a reflection of the lack of fully developed operational techniques within economies for dealing with market failure issues but it is also the result of the lack of political will by governments to take concerted action. Moreover, the issues are considered too fundamental and far-reaching to be within the empowerment of tourism firms. This is exemplified in developed countries but is even more apparent with respect to many tourist-emergent developing countries where access and property rights regarding resources are obscure or non-existent, government legislation and regulations are absent, ill-considered or badly enforced and there are no appropriate institu-tional structures. None the less, the concept of market failure has been important in generating a body of research on the estimation of use and non-use demand, the evaluation of that demand and the derivation of prac-tical policy instruments to mitigate the adverse effects of markets or to increase their efficiency in both developed and developing countries. This research is considered, in relation to tourism, in Chapter 8, in which the examination is focused on economic methods rather than the general con-cepts and issues which have been discussed above.

8

ENVIRONMENTAL VALUATION AND SUSTAINABILITY

INTRODUCTION

The previous chapter was concerned largely with the wider environmental issues of economic impact in general and tourism in particular and their implications for sustainability and the conservation of exhaustible and renewable productive and energy resources. It also introduced the concept of market failure and its consequences for the environment, especially the natural resource base. Furthermore, the chapter emphasized the importance of the analytical framework of cost-benefit analysis and its application to environmental problems. Fundamentally, therefore, attention was concentrated on demonstrating the relevance of economic principles to appraising the environmental impact of tourism activity.

In this chapter economic methods of identifying and evaluating the environmental benefits and costs of tourism are examined in some detail and the policy instruments which can be applied to enhance the former and mitigate the latter are evaluated. The relative advantages and disadvantages of regulatory as opposed to price or market-based instruments are then considered. The chapter is concluded by offering some observations on the state of the art of economic methods and prescriptions for environmental action. A number of desiderata as necessary steps to pursue sustainable tourism are identified and comments on the likelihood of environmental initiatives by firms are made.

The chapter starts from the point in Figure 7.1 (pp. 162–3) presented in Chapter 7, which concerns the valuation of resources following on from the problems associated with market failure. There is almost a sequential relationship between the three main forms of market failure discussed in Chapter 7 and identified in Figure 7.1 and the remainder of the figure which concerns the concept of economic value, valuation techniques and policy instruments.

THE VALUATION OF RESOURCES

The problems arising from the existence of public goods, the generation of externalities and the need to consider distributional issues are founded in the need to evaluate demand in order to make decisions on the allocation of resources. Demand evaluation, or perhaps more correctly the estimation of consumer benefits, is necessary irrespective of whether or not demand is expressed through the market. Except under specific conditions, economics accepts that the market price does not necessarily represent the value of a good or service and clearly for non-traded commodities, where no price exists, does not suggest a zero value. In the latter case, means have to be devised to attach value. There are thus two elements to estimating consumer benefits. The first is to establish what is meant by value in use and non-use. The second is to employ techniques to ascertain that value.

Total economic value

Although the non-priced elements of the environment have been empha-sized, it must not be forgotten that a number of environmental goods and services are traded in the market where excludability, i.e. private consump-tion rights, can normally be exercised. In the market, the exchange value is indicated by the price at which it is traded. Nevertheless, in economics it is recognized that the use value can be greater than this price for all but the marginal consumer, on the assumption of a downward sloping demand curve from left to right, as there are many purchasers who are willing to pay a price above that which prevails in the market. This is known as consumers' surplus, which yields an aggregate use value above the price paid by individual consumers and the quantity each purchases. In practice the concept of consumers' surplus is rarely invoked for priced goods to ascertain total user value, but it is of interest because it forms the theoretical foundation of a valuation method to establish the willingness to pay for non-priced goods.

Although environmental benefits are often unmeasured and unpriced, the true value of many collective consumption goods can be considered as being much greater because they are unique, so that if they are over-used an irre-versible trend may be initiated which leads to their destruction. Moreover, as unique resources they cannot be reproduced. This often imparts a non-use or passive value in addition to the use benefits. Thus they have a value which transcends their exchange and use value, i.e. respectively, any price paid and the consumers' surplus. In the context of environmental and collective con-sumption goods, a number of non-use benefits or values have been identi-fied which should be added to the use values to yield what is known as total economic value (TEV).

With respect to static benefits, which are generated at one point in time and arise from existing resources, such as heritage sites and public spaces, in effect from the stock of resources in their present state, it is possible to per-ceive of the natural and human-made environment as possessing total eco-nomic value consisting of use value and non-use value.

183

Use value

This is both direct and indirect, for example the occupation and use of an historic building, whether purchased or rented, would represent direct value which may include consumers' surplus. Its appearance, giving pleasure to occupiers, the local community and tourists, would give rise to indirect value, constituting an unpriced externality and therefore necessitating the estimation of the benefits to arrive at a total use value.

Non-use value

Economists suggest that there are two main forms of this demand: option value and intrinsic or existence value.

Option value

This is the potential benefit which consumers might derive from resources. It is an expression of a willingness to pay for their preservation so that they retain the possibility of using them in the future. In this sense, option demand is a quasi-use value. It may be extended to include an option for others to enjoy the consumption of certain resources – a kind of vicarious demand. Some economists distinguish between demand by the current and future generation. The term 'bequest value' has been coined to suggest the value which the present generation places on resources, where it expresses a willingness to pay for their preservation for the benefit of future generations. This, however, can be construed as a form of option demand and is viewed as such here. For example, potential tourists might be willing to pay to prevent an historic building from being demolished or a forest felled or a beautiful coastline from being developed.

Intrinsic or existence value

This is a more complex and unclear form of value in that it can be considered as unrelated to demand. People may have preferences for and therefore place value on the continued existence of resources, with no intention of ever using them. Therefore the preservation of both natural and human-made resources may be advocated because it is recognized that they have intrinsic value. Individuals may be willing to pay simply to know that an area or building would be conserved even though they expect never to visit it. This is mainly relevant in the case of world land marks (the rain forest or Grand Canyon, for example) but may also be relevant for whole historic city centres/public urban spaces (Oxford, Paris or Venice).

Therefore TEV = Use value + Non-use value consisting of option value + existence value

So far the analysis has been based on the implicit assumption that there is a given stock of resources and environment, so that the question is what value do people attach to this? Another approach is to assume that only one variable changes in order to observe changes in the value placed on resources, for example, how can the possibility that future generations may have different tastes be taken into account? In addition to these static environmental benefits there are dynamic ones, that may be inadequately measured in market prices or which market prices do not reflect at all. It is no longer just a question of estimating the size of a given stock; it is the possibility that action may be taken which acts as a catalyst in making the stock larger. For instance, many human-made resources, such as the creation of lakes, enhancement of the landscape and improvement of the built environment, also produce indirect benefits if people are attracted to live in or visit such areas. This argument derives ultimately from a seminal contribution of O.A. Davis and Whinston (1961). While they were concerned with strategies by householders to under-invest in their properties, acting as free riders and so leading to the creation of slums, their argument can be considered in reverse whereby intervention by public agencies to conserve and improve the environment will result in the enhancement of, say, land values and rents. This argument will be considered below in the context of tourism.

The total economic value of tourism

While the notion of total economic value (TEV) is the foundation of the estimation of the benefits derived from resources in environmental economics, its relevance to evaluating the tourism resource base has hardly crept into the literature. However, given tourism's reliance on the natural environment it is not difficult to appreciate its implications and application. With regard to its implications, TEV suggests that the resource base, which is treated largely as a free good by tourism firms, has a much greater value than has been envisaged. This should increase the importance of the environment as an input into the tourism product by raising its opportunity cost in terms of alternative uses in, for example, agriculture, industry or urban property development. It should also emphasize to governments, tourism bodies and the business sector the necessity of safeguarding this resource base.

The application of TEV to tourism, bringing out the distinction between static and dynamic benefits, is demonstrated by an example of the role of heritage conservation and tourism in urban economies, particularly the regeneration of post-industrial towns and cities. Attention is concentrated on the built environment. Cities, it is argued, have ceased to have an important function in the direct production and distribution of goods. Rather they are reverting to the functions that they had before the industrial revolution as sites for administrative services business, and commercial activity, as generators of cultural services and, increasingly, as the providers of urban amenities, both enhancing the quality of life of their residents and providing the economic base for urban tourism and other leisure-oriented service activities. The key difference with the pre-industrial era is that the set of activities identified above now contributes a growing part of urban economic output and welfare.

The likelihood of an urban area succeeding in changing its function in this way depends, in part, on its setting, culture and legacy of historic buildings and areas. If it can both make the best use of its inherited assets and change its function, intervention will assist the regeneration of the local economy. By investing in conservation policies for architecturally significant buildings or areas, public agencies may encourage owners of adjoining properties to upgrade them. People with new skills can be encouraged to move into those areas or skilled people persuaded to remain there. This could lead to new businesses being located in the conserved buildings/areas and, in turn, facilitate the emergence of new business opportunities, additional spending and engender yet more new business creation. In effect, the implication is that public spending fills a gap that the market, left to itself, would not: O.A. Davis and Whinston's (1961) 'neighbourhood effect'. The initial public investment also acts to give leverage to further private investment, both in upgrading other buildings in the area and in the creation of new businesses or the retention of businesses that would otherwise have failed or left. However, this is only the medium-term effect. The argument can be extended to the longer-term process of cumulative upgrading as the import of new skills reinforces the base of local entrepreneurial activity and improves the image of the area. Such a process is akin to multiplier-accelerator analysis which has been extensively considered in relation to regional development (Richardson, 1972) and investigated with respect to the role of specific economic activities, particularly tourism (for example, Sinclair and Sutcliffe, 1988a; Archer, 1989; Johnson and Thomas, 1990).

Thus, while the static benefits of the impact of an improved urban environment can be measured by means of 'one-off' studies, reflecting the TEV concept, by estimating property values and/or willingness to pay to live or work in or visit a city or town (the valuation methods are reviewed in the next section), the assessment of the dynamic benefits has to be attempted by observing and evaluating the process of change. This is more difficult given the techniques derived so far in environmental economics. It therefore constitutes an issue which needs to be addressed if the static benefit analysis is to be developed to accommodate the dynamic nature of economic activity and the consequent variation in benefits over time. Moreover, it is imperative that net social benefits are measured, i.e. that the social costs of changes are taken into account. Initially a case-study approach seems feasible. The measurement of static and dynamic benefits will be discussed further after an appraisal of environmental economic valuation methods.

VALUATION TECHNIQUES

In the development of the analysis of the amenity and pollution aspects of environmental economics, much attention has been paid to their non-traded nature. The economics literature addressing the issue of how to value environmental deterioration or improvement or the amount consumers would be willing to pay to preserve, say, an unspoilt beach, forest, historic building or undeveloped landscape of high scenic value, is vast. The main methods which have been developed to place a value on environmental attributes and which could be applied to the valuation of non-priced tourism resources are hedonic pricing, travel cost, contingent valuation or combinations of these three. Of possible value where detrimental impacts occur are those rooted in household production function analysis such as avoided costs and estimation of the opportunity cost, provision or replacement cost. They can be classified according to whether they seek to place a value on the good or attribute directly, by asking respondents their willingness to pay for an improvement or their willingness to accept a degradation, or indirectly, by using prices from a related market which does exist. Contingent valuation is an example of the former, while hedonic pricing and travel cost are representative of the latter. In addition there are more qualitative approaches, such as the Delphi technique, a direct method, which might be employed. The three most commonly applied methods are appraised first as being most relevant to the valuation of tourism resources. Indirect techniques are based on the concept of consumers' surplus in that, whether or not a price is paid for a specific resource, a measure of the net benefit is estimated.

The hedonic pricing method

The hedonic pricing method (HPM) developed by Rosen (1974), based on earlier consumer theory of Lancaster (1966) and Freeman (1979), can be used to estimate the value of unpriced characteristics of goods and services. It aims to determine the relationship between the attributes of a good and its price and is arguably the most theoretically rigorous of the valuation methods. It takes as its starting point that any differentiated product can be viewed as a bundle of characteristics, each with its own implicit or shadow price. For example, in the case of housing where it has been widely applied, the characteristics may be structural, such as the number of bedrooms, size of plot, presence or absence of a garage, or environmental, for instance air quality, the presence of views, noise levels, crime rate, proximity to shops or schools. Accordingly the difference in price between two houses, identical in every respect except for one, should accurately reflect consumers' valuation of that characteristic. Likewise, it is possible to attribute the impact of the designation of sites as amenity or conservation areas, which attract tourists to historic cities and towns, by observing the difference in value between two identical sites, one of which is regarded as a conservation area and the other of which is not. Thus the price of a given resource, for example the destination environment, can be viewed as the sum of the shadow prices of its characteristics.

Most studies using the HPM have been concerned with the impact of amenity resources, such as those which have considered the effect of the proximity of environmental and neighbourhood variables on residential property prices, for example, a forest or nature reserve. They are relevant to tourism in that the methodology may also be used to estimate the effect of such variables on holiday prices, as well as identifying and measuring the environmental impact of tourism. Garrod and Willis (1991a, 1991b, 1991c) found that the presence of forestry may have a large positive impact. A similar result is observed in the case of location near waterways. Willis and Garrod (1993a) estimated that the presence of a canal or river raises the value of a property by an average of 4.9 per cent, while the proximity of at least 20 per cent woodland cover raises it by 7.1 per cent above that of an identical property without these features. Cheshire and Sheppard (1995) investigated how the value of location-specific attributes are capitalized into land prices if they are not included as independent variables. Among the variables included were local amenities provided through the land-use planning system.

There also exists a significant body of research, with possible relevance to tourism in historic urban areas, into the impact of architectural style and historic zone designation on property valuation. Asabere et al. (1989), for example, showed that architectural style has a strong impact on residential property valuation in their sample of 500 properties sold in Newport, Massachusetts, USA, between 1983 and 1985, with older styles of architecture commanding premiums of around 20 per cent. Moorhouse and Smith (1994), in their study of nineteenth-century terraced houses in Boston, also found that individuality of any style commanded a higher price. Hough and Kratz (1983) argued that architecture has certain public good characteristics which may be undervalued in the market while Ford (1989) evaluated the effect of Historic District designation on the prices of properties sold in Baltimore, Maryland, USA, between 1980 and 1985. Designation was found to have a positive but insignificant impact. This result was corroborated by Asabere et al. (1989) and also by Schaeffer and Millerick (1991) in their study of the prices of 252 properties prior to and after designation in Chicago.

A related area of study in which the HPM has been used is in research on the pricing of package holidays where environmental characteristics, in effect the tourism resource base, are part of the tourism package. The price competitiveness of package holidays cannot be compared directly because of differences in the characteristics of the packages supplied. However, the HPM can be used to estimate the price differences which are due to variations in the mixes of characteristics. For instance, in the case of holidays in the

Spanish province of Malaga, offered by UK tour operators, the characteristics included the categories of hotels in which accommodation was supplied, the hotel facilities and locations and the tourist resorts themselves, all of which affected the prices of the holidays (Sinclair *et al.*, 1990). The HPM has also been used to compare the price competitiveness of package holidays in cities, supplied by tour operators from different countries (Clewer *et al.*, 1992).

The travel cost method

The travel cost method (TCM) was developed by Clawson and Knetsch (1966), following an initial suggestion of the plausibility of the technique by Hotelling (1949). The method is based on the premise that the cost of travel to recreational sites or tourist areas can be used as a measure of visitors' willingness to pay and thus their valuation of those sites; it has been applied to the demand for boating, fishing, forest visits and hunting. The costs of travelling to a site are taken as a proxy for the price of the product. Thus, even if visitors do not pay to use a site, they have incurred expenditure either implicitly or explicitly in travelling to it, which could be used as a measure of (or at least a lower bound to) their valuation of that site. Time can be perceived as an implicit cost while explicit costs include petrol or public transport fares. Whether on-site and travelling time should be incorporated into the estimate of total cost is a point of debate in the literature (V.K. Smith and Desvouges, 1986; Chevas *et al.*, 1989). For sites to which the majority of visitors walk, valuing the time they take is the only measure which can feasibly be used (A.J.M. Harrison and Stabler, 1981). This might suggest that on-site time should be the true focus of the debate. If it is decided to include on-site travelling time it may be difficult to assign a value to it. The opportunity cost in terms of forgone earnings or leisure time which could have been spent doing something else might be taken.

Studies employing the travel cost method almost exclusively consist of those for visits to rural recreational sites such as important areas of high scenic value, forests, lakes, mountains or water courses. These applications can be categorized according to whether they use an individual, zonal or hedonic formulation of the travel cost model, or a combination of all three. Englin and Mendelsohn (1991), for example, estimated the value of alterations in the quality of forest sites using a hedonic travel cost model. They found that some site attributes, such as dirt tracks and alpine fir trees, had saturation levels, below which they are an economic good but above which they are a bad. Hanley and Ruffell (1992) used the travel cost method to evaluate consumer surplus across different types of forestry, each with different physical characteristics. The study showed a strong relationship between visits per year and the mean height of trees, reason for visit, length of stay and importance of visit, while forest characteristics were insignificant. In a study of canals (Willis *et al.,* 1990) and another of botanic gardens (Garrod *et al.,* 1991), consumer surplus valuations were estimated to be significantly lower than the financial operating loss. Applications to the valuation of the quality of fishing (V.K. Smith *et al.,* 1991) and deer-hunting sites (Loomis *et al.,* 1991) have also been considered. Very few studies using TCM have been conducted to evaluate urban resources, two principal reasons being the problems of separately identifying the benefits generated by specific resources and the many reasons those travelling to cities and towns have for their visits.

The contingent valuation method

In contrast to the hedonic pricing and the travel cost methods, which are indirect methods of eliciting valuations from consumers by considering their revealed consumption in related markets, the contingent valuation method (CVM) directly questions consumers on their stated willingness to pay (WTP) for, say, an environmental improvement or their willingness to accept (WTA) compensation for a fall in the quality of the environment. Since respondents are questioned directly, it is possible to ask them whether they would be willing to pay, for example, to preserve a recreational site or even a tropical rain forest of which they are not users. Thus an advantage of the method over others is that it is possible to obtain, at least in principle, non-use option and existence valuations as well as use values. A standard text on the method is R.C. Mitchell and Carson (1989) and significant contributions have been made by Brookshire *et al.* (1983), Desvouges *et al.* (1983), Heberlein and Bishop (1986) and Hanley (1988).

The CVM is a survey-based methodology with responses being obtained either by questionnaire or by interview. A number of stages are involved in implementing the survey, the first of which is to set up a hypothetical reason for payment or compensation in the eyes of consumers. So, for example, respondents may be told that the government is considering clearing and improving a nearby former industrial site for a leisure development, but only if additional funds are raised. The respondent is then informed about how much additional revenue is required and how the scheme would be paid for, i.e. the 'bid vehicle', for example a local income tax or an entry fee. Individuals are then questioned by the various means which have been developed within the method on their maximum WTP to ensure the scheme goes ahead, or on their minimum WTA compensation for loss of the resource or artefact. The final steps in the analysis involve estimation of a bid curve involving a regression of the WTP/WTA on a range of explanatory variables which are thought to affect the bid and on which information has been elicited in the survey, for example incomes and educational levels.

A rapid growth in the number of applications of some form of CVM has taken place since the mid-1980s. This may be partly attributed to the potential ability of the method to value option and existence values. In the UK, the option value to preserve existing landscape was found to be between 10 and 20 per cent of the total site valuation (Willis, 1989; Bateman *et al.*, 1994). Lockwood *et al.* (1993) used a CVM survey in order to assess WTP to preserve national parks in Victoria, Australia. The survey highlighted the relative importance of existence and bequest values, which constituted 35 per cent and 36 per cent of the total valuation respectively. Non-use value was found to be over three times that of use value in a survey of the Somerset Levels and moors Environmentally Sensitive Area (ESA) scheme (Garrod *et al.*, 1994). Other related areas of application of contingent valuation have included improved park facilities (Combs *et al.*, 1993), a ban on the burning of straw (Hanley, 1988), the benefits of canals (Willis and Garrod, 1993a), forestry characteristics (Hanley and Ruffell, 1992, 1993), wildlife (Willis and Garrod, 1993b), the value of elephants (Brown and Henry, 1989) and tourism-related traffic congestion (Lindberg and Johnson, 1997). In a paper related to the evaluation of urban resources in the UK, Willis *et al.* (1993) assessed the usefulness of CVM for estimating willingness to pay to gain access to historic buildings. They considered CVM to be a constant, robust and efficient estimator of WTP, although non-use value was not measured in their particular study. A qualified endorsement of CVM has been made by the National Oceanic and Atmospheric Administration (NOAA, 1993) using what amounts to a Delphi technique, which is examined on p. 194.

OTHER METHODS OF VALUATION

Although, together, HPM, TCM and CVM account for the vast majority of practical applications, other methods of valuation exist.

Production function method

The basis of the production function method (PFM) is that firms, households and individuals combine factors of production and commodities respectively with environmental services in order to produce other goods and services, for example, their holidays. This acknowledges the complexity of demand for leisure pursuits and tourism which involves several components, some of which may have to be combined sequentially and others simultaneously. The method is potentially useful as it allows 'before' and 'after' situations to be assessed so that if the quality or quantity of an environmental attribute changes, agents change their expenditure patterns in order to take account of this change in the environmental attribute. The problem with PFM is that it has been difficult to make it operational in what might be termed its pure form. Many authors (for example, Bateman *et al.*, 1992) argue that PFM forms the basis of other approaches, such as TCM, dose response and avoided cost (sometimes referred to as averted expenditure).

Dose–response functions

These have generally been applied in the context of pollution to estimate the effect of changes in environmental quality on crop yields. For example, a change in the levels of pollution present may cause farm production to alter, or air pollution damage to buildings might affect their utility to occupiers or impair their visual appearance and so deter visits by tourists. Thus the financial effect of changes in the quality of environmental services can be measured and quantified.

Avoided cost method

The basis of avoided cost methods is that economic agents may be able to undertake an expenditure to minimize the effect of a fall in environmental quality. For example, if the quality of drinking water is considered to have fallen below an acceptable standard in a tourist destination, the local authorities may decide to install a water filter system, the cost to be borne by local suppliers. The sum of these expenditures across all affected parties could be viewed as an implicit valuation of (a lower bound to) the fall in water quality. In a study of direct expenditure by consumers, Hansen and Hallam (1991), for example, use a production function approach in the valuation of recreational fishing benefits of an improvement in freshwater stream-flow relative to 'consumptive' river uses, such as agricultural irrigation. They found that marginal valuations of stream-flow changes are higher for recreational fishing than for their corresponding use in agriculture. Clearly, however, it is likely that the averting expenditure and the environmental service forgone are imperfect substitutes. For example, many tourists experience a loss of welfare as a consequence of the environmental degradation in general as well as of unique sites of high scenic value, but do not consider the averting expenditure (which may have a high threshold) worthwhile; for instance the use of public transport at a higher personal cost than using a car to reach a holiday destination or a site for which the visual intrusion of parked vehicles adversely affects its attractiveness. Thus valuations using avoided cost methods may produce underestimates and this has been found to be the case in many human life valuations (Dardin, 1980; Dickie and Gerking, 1991). There is, however, no specific evidence of tourists being directly unwilling to pay to avoid being exposed to risky environments which might endanger health and life.

The relevance of these methods to conservation and natural resources is that they potentially can measure the impact on welfare of ongoing changes in the environment, i.e. dynamic benefits and costs. The examples cited from the literature largely refer to degradation of the environment and being applied in this way they could indicate the impact of a 'without' conservation situation. However, in principle there is no reason why they cannot measure the impact of an improvement. Their main shortcoming is the difficulty of making them operational.

Opportunity cost

Estimating the value of resources used for a specific activity in terms of the opportunity cost is a standard concept in economics but is seldom applied in practice because of the difficulty of establishing an acceptable measure. It is possible to ascertain the difference in value between the use of resources without any constraints as opposed to their use with them, because of environmental objectives. For example, the constraints imposed on the ownership of an historic building by being listed and its subsequent maintenance are likely to reduce the market value or rents derived from it. The difference can be taken as an indication of its heritage value. The problem is that this is almost certainly an under-valuation as it regards conservation in purely market terms because non-use benefits and externalities are ignored, such as the benefits accruing to tourists wishing to look at conserved buildings. A possible resolution of this omission is referred to in discussing replacement cost.

Replacement cost

Replacement cost is a variant of opportunity cost as an accepted economic approach to ascertaining the value of resources. It differs because it does not consider alternative uses of a specific resource but the cost of providing a similar resource as a substitute. This cost can represent the value of the original resource. There are two problems with this method of valuation. The first is that it is suitable where resources are easily reproducible, for instance an artificial boating lake, but is inappropriate for unique natural environments. Second, it suffers from the same shortcoming as the opportunity cost approach in that non-use benefits and externalities would normally not be included, if market-based replacement cost is considered. This problem can be overcome by applying one of the methods for estimating total economic value, such as CVM, in which case the whole valuation exercise might as well be conducted in this way, thus making the calculation of replacement cost redundant.

Provision cost

Early studies by suppliers of the benefits derived from leisure resources considered that their provision expenditure could be taken as a measure of the value of such resources. While such expenditure may reflect society's willingness to pay for amenity and environmental goods, the approach is patently illogical. Providers, by merely spending more on such projects, can claim that they have greater value. This strategy might be adopted to achieve the objectives of the providing organization. It is very unlikely that valuation undertaken on this basis will correspond with that which would be stated or revealed by consumers and, as with the opportunity cost and replacement cost methods, would certainly not measure total economic value.

Delphi technique

Instead of attempting to ascertain the WTP for an environmental improvement or WTA for a degradation of it by individuals who may be irrational, impulsive or poorly informed, it is possible to employ a panel of 'experts' and to elicit their views on the valuation of environmental changes. The technique was developed by the RAND Corporation in the 1950s (Dalkey and Helmer, 1963) and has been found to be particularly useful in cases where historical data are unavailable or where significant levels of subjective judgement are necessary (S.L.J. Smith, 1989). The method consists of the researcher assembling a panel of those who are believed to have some knowledge concerning the issue at hand. It is important that the panel be from diverse fields, with different approaches, and therefore different viewpoints and subjective valuations. The numbers in panels have ranged anywhere from four (Brockhoff, 1975) to over 900 (Shafer et al., 1974). In the next stage the panel is given information on the study and asked for individual valuations. The responses are circulated to all panel members, who are asked if they wish to revise their own valuation in the light of the other responses. This process continues until some convergence of views between the experts, in the light of discussions with their colleagues, has occurred. An advantage of the method is that it requires little specialist statistical knowledge and is relatively simple to conduct, although the selection of an appropriate panel, and the design and wording of the questionnaire, are likely to have a crucial influence on the final outcome. Some may also view the process of valuation by so-called 'experts' as undemocratic and artificial.

In discussing the environmental impact of tourism and means of assessing it, Hunter and Green (1995), in addition to canvassing expert opinions, emphasize the necessity for the local community to be involved in the decision-making process and the valuable contribution it can make to directing or assisting the investigation of potentially significant impacts. Korca (1991) acknowledged the role of experts and local residents in using the Delphi technique in a study of tourism development in the Mediterranean, in which a panel composed of representatives of both groups was assembled to identify the impact of continuing to expand tourism. Green and Hunter (1992) incorporated local public opinion when using the technique to assess redevelopment at a site in northern Britain. Green et al. (1990b) also considered the more general application of the Delphi technique to tourism.

THE ATTRIBUTES OF THE THREE MAIN METHODS AND THEIR APPLICATION TO TOURISM

A number of possible methods of valuing non-market goods have been examined, almost all of which have been applied in a leisure or environmental quality context. They are equally relevant to tourism, given the role of the environment as its resource base. Their principal relative strengths and weaknesses have been identified and it appears that there may be some trade-off between economic rigour and breadth of applicability.

The hedonic pricing method (HPM) takes account of the value that consumers place on environmental attributes only in so far as they enter into product prices or are capitalized into land and property values; i.e. it captures internalized benefits. Its validity in the context of environmental factors in tourism therefore depends on the extent to which the price of the tourism product is determined by the attributes of its environmental components, for example the quality of the beaches or facilities or attractiveness and cultural associations of historic buildings. Even if such environmental attributes are embodied in market prices the technique is neither suitable for the consideration of option, existence or bequest values, nor does it incorporate non-market benefits enjoyed by visitors. Nevertheless, its advocates assert that it is acceptable from an economic standpoint because of its rigour, reliability and robustness.

A number of problems with the use of the HPM as a technique for the valuation of non-market goods and, in the present context, of conservation, have been described in the literature, in addition to it capturing only benefits which manifest themselves in market prices, resulting in the potential for gross undervaluations of environmental or historic conservation. The HPM framework is based on a number of restrictive assumptions including consumers who are perfectly informed with respect to prices and attributes at all possible locations, and a market equilibrium. If these are violated, it is possible that valuation based on the HPM may be inaccurate to an unknown extent. Data requirements may also be onerous.

Further difficulties arise with regard to the estimation of the hedonic equation. These include bias from omitted but relevant variables which might distort the estimated valuations, an inappropriate functional form for the demand curve, the presence of multicollinearity (closely related variables) and the presence of market segmentation which is ignored or overlooked, so giving biased results. For example, even within seemingly homogeneous tourism markets, such as those for packaged summer sun holidays, there is segmentation by such variables as price, quality, destination and accessibility. The HPM must acknowledge such segmentation and estimate separate equations for each segment. These are viewed as serious drawbacks at a technical level and impair the confidence that can be placed in the results.

Willis *et al.* (1993) argue that the travel cost method (TCM) works best when the majority of the visitors live at a large distance from the site. Hence TCM models are appropriate for measuring the benefits of tourism but may be inadequate for the valuation of urban recreational sites where the distance travelled is likely to be very short and so tends to nullify the technique. However, the TCM may be relevant in these cases if the value of travel time is incorporated so that even those living nearby who walk to the recreational site incur a cost. The TCM is appropriate for both resident tourists who use the site, i.e. those accommodated in the locality, and for visitors from further afield. However, if former tourists are so attracted to an area that they move to live nearer to sites in order to visit them more often, the TCM will under-value benefits because travel costs will be lower.

Both domestic and international tourists pose a fundamental conceptual problem for the TCM because the distance travelled to attractions should not include that from their point of origin for the holiday, unless the sole purpose is to visit a specific attraction. If tourists stay in one location from which they travel to sites then there is no problem; the distance for each trip can be employed. However, holidaymakers who are touring and visit a number of sites on one outing raise the question of the cost of travel to each site (Cheshire and Stabler, 1976); a method of apportioning the travel costs to each site has to be devised and may be arbitrary. The TCM is perhaps the least preferred of the three main methods technically, since it has neither the strong economic theoretical underpinning (the consumers' surplus concept is contested) nor the well-developed economic theory of hedonic pricing. It does not possess the attributes to measure wider non-use values and it is dif-ficult to aggregate values obtained for individual sites.

There is an extensive critical literature on the TCM, particularly with regard to the problems associated with its statistical attributes. For example, there are two approaches as to the choice of the dependent variable of the num-ber of visits: the visit rate from given distance zones or the number of visits made by given individuals (Willis and Garrod, 1991a). Another problem is that there is no consensus as to the preferred functional form in the case where the demand curve is asymptotic (does not intersect the axes) (Common, 1973; Hanley, 1989). This occurs if there are extreme outliers either in the number of visits or the distance travelled and hence the travel cost incurred. This raises the issue of whether or where to use cut-off points. Since, by definition, an on-site survey would not capture the value placed on it by non-users or by those who do not make a visit during the sample period, an under-estimation of consumer surplus would occur (V.K. Smith and Desvouges, 1986). Both the HPM and the TCM suffer from requiring large resources in terms of time and funds to collect, process and analyse data. They are also suspect because of the missing variable, multicollinearity and functional form problems.

The contingent valuation method (CVM) is possibly the least rigorous but arguably now the most popular of the three main valuation techniques. It has become widely accepted by politicians since it is compatible with the concept of democracy and valuation by the population at large. It has the merit of being not only simple, flexible and easily understood but also the broadest in covering use and non-use values and both residents and visitors. Its main drawback is that, to date, it has investigated virtually only hypothetical situa-tions. However, it could be applied to real-life circumstances where a decision has to be made on specific resources when WTP and/or WTA need to be ascertained and payment and/or compensation made. It is now more readily accepted, particularly in the USA, as a valid basis for compensation in cases of environmental degradation. However, despite a substantial body of academic work on its use with respect to environmental quality and pollution, it has as yet not been widely applied as an appropriate method of establishing society's evaluation of natural and built environments and a feasible means for making decisions on resource allocation.

The CVM is certainly the least technically developed of the valuation methods considered thus far. An extensive review of the many biases or errors which may be inherent in CVM studies is given in Garrod and Willis (1990). Their nature and significance is not examined here in detail but some of the most important are identified and ways of overcoming them outlined. Strategic bias results from 'free riding', that is the deliberate misrepresentation of respondents' own WTP/WTA in order to serve their own interest. Thus if consumers think that their payment could be based upon the valuation they state in the survey, they have an incentive to understate their true valuation if they think the improvement will go ahead irrespective of their valuation. This problem can be overcome by stating that all consumers will pay the average bid. Hypothetical bias occurs when it is plausible that responses to questions on a hypothetical situation differ considerably from those which refer to a real event. A mis-valuation may occur if the respondent finds the situation difficult to relate to, or simply fails to take the survey seriously, therefore giving ill-thought-out responses. The researchers therefore need to make the issue and means of dealing with it appear as realistic as possible.

Bias is apparent when the respondent appears to value large differences in environmental quality almost identically. Thus if individuals are asked to take part in a number of CVM surveys, it is possible that their total WTP would be a large portion of, or even exceed, income. When asked about one specific issue, individuals often allocate their whole 'environmental budget' to it since they are unaware of other possible environmental improvements to which they may be willing to contribute (Willis and Garrod, 1991b, 1991c). It is therefore necessary to try to establish the total environmental perception of individuals. When a respondent appears unable to distinguish between, on the one hand, a WTA payment to be denied one visit to a site and, on the other, a WTA payment for permanent removal of the possibility of ever visiting the site, then what is termed the temporal embedding problem arises.

A number of biases may result from the way in which the questions are phrased and the survey is implemented. Starting-point bias occurs when the starting value suggested to the respondent under the iterative bidding technique, if it alters the final bid offered, is viewed as the sum which should be the WTP or the average or expected bid. The method of payment proposed in the survey may also influence the WTP/WTA figure given. For example, it has been found in a number of applications that WTP via a sales or income tax was greater than that via an entrance fee. Hanley and Spash (1993) suggest the use of non-controversial payment systems which are most likely to be used in practice to improve credibility.

ACHIEVING ENVIRONMENTAL TARGETS: POLICY INSTRUMENTS

The economics literature tends to be dominated by the consideration of two aspects of the pursuit of environmental goals. The first is the advocacy of the price mechanism. The second is the tendency to concentrate on mitigating the effects of environmental degradation, particularly pollution. Indeed, in the political and social consciousness of environmental problems, the utterance of the words 'polluter must pay' has become a catch-phrase for action. This is misleading for not only does it imply that the sole environmental issue is pollution but it infers that those economic agents, such as primary and secondary producers, are the principal sources of environmental problems. It also suggests that the costs of dealing with these problems should fall entirely on the perpetrators. More holistically thinking environmental economists do not take this somewhat oversimplistic view. They recognize that as well as social costs there are many instances of benefits which are bestowed on society at large, for example by owners of natural resources and by businesses, public bodies and individuals in their normal day-to-day activities, for which no payment is made by users. This branch of the subject also accepts that any intervention to correct market failure or imperfections may itself distort the distribution of welfare and allocation of resources; indeed the form of intervention, i.e. the instruments employed, can have unintended effects. Therefore, it seems appropriate to consider together both the nature of the instruments and their evaluation, even though these have been identified as separate issues in Figure 7.1.

Economists largely favour the price mechanism because it tends to lower the costs of dealing with environmental problems, to make the cost of production and consumption more visible and to offer incentives and disincentives in the market place, potentially minimizing distortions. They also acknowledge that some non-monetary instruments play a role in circumstances where price-based instruments might prove inappropriate or where they are complementary, or have an impact on prices and costs. This explains the presence in Figure 7.1 of non-market instruments, which in practice tend to predominate in industrial countries. It is appropriate to make a further distinction between once-and-for-all instruments, for example those concerning capital investment, and those which are recurrent.

In examining the various instruments which have been proposed in the environmental economics literature, attention is concentrated on the issues each raises rather than simply describing their purpose and how they are applied. A fundamental debate concerns the relative merits and demerits of price-based and regulatory (sometimes designated command and control) methods.

PRICE-BASED INSTRUMENTS

Price-based instruments for dealing with environmental issues take two forms. One, for example, a tax or subsidy, attempts to modify the operation of existing markets whereas the other, such as tradable quotas or permits, tries to create markets which did not previously exist. The economic argument for using the price mechanism stems from the recognition that the cost structures of different activities vary and so in contrast with regulation which imposes a standard compliance, there can be more flexibility in the response by those subject to price-based environmental policies. For instance, in the case of a tax on air emissions, a firm can elect to pay the tax or alternatively fit control equipment, the response being determined by the relative costs of the tax instrument and the cost of investment in equipment and/or procedures to reduce or eliminate the environmentally damaging activity. The reverse could apply with respect to conferring unpriced benefits whereby a subsidy would be preferred by the supplier if it exceeded the cost of capturing formerly unpriced benefits via the market. For example, many landowners of natural resources incur costs in making them available to tourists, yet cannot exclude visitors, perhaps because the cost of fencing and/or employing people to collect an entry fee would outweigh the revenue generated; a local authority may pay a subsidy to ensure the resource remains available. The subsidy might relate to management of the resource as an amenity or compensation for damage or loss of revenue from an alternative productive use, such as crops or timber. Another factor of some importance is the effectiveness of price-based instruments with respect to the burden on the public purse. Regulation through the price mechanism may lower enforcement costs; in effect, once the price instruments are implemented and accepted, there can be virtual self-regulation by agents active in the market. The principle underpinning price-based instruments is that the full production or consumption costs or benefits should be estimated in order that the appropriate level of charge, tax, subsidy or grant can be ascertained to achieve a socially optimal position where the marginal social benefits and costs are equal, i.e. the full costs or benefits are reflected in market prices as shown in Figure 8.1.

Figure 8.1 is essentially a simplification of Figure 7.2 and indicates why an economic optimum position is gained at Q_S. The figure shows the private optimum at Q_P where marginal private cost (MPC) equals the marginal private and social benefit (MPB and MSB, assumed equal, for simplicity). At the activity level Q_P and price PP in a market economy which does not take account of environmental costs, the marginal social costs (MSC) are ignored. Consequently the price does not reflect the full costs to society of the production of

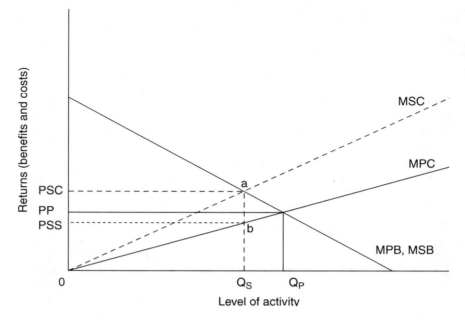

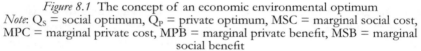

Figure 8.1 The concept of an economic environmental optimum
Note: Q_S = social optimum, Q_P = private optimum, MSC = marginal social cost, MPC = marginal private cost, MPB = marginal private benefit, MSB = marginal social benefit

the good or service and its consumption. By levying a charge or tax equal to the difference (ab) between the MPC and MSC at Q_S, an economic optimum is achieved, i.e. where MSB equals MSC. Reduction of the level of activity to the left of Q_S would be economically inefficient as the loss in benefit outweighs the reduction in MSC, i.e. MSB is greater than MSC. It is worth noting that consumers as well as suppliers bear some of the burden of a charge or tax on suppliers as market price increases to PSC after its imposition; suppliers also bear part as their net of tax price is PSS as opposed to PP previously.

In the context of tourism a landing fee or bed tax has the effect explained above and in Figure 8.1 if it is levied on suppliers in the first instance. A proportion of the charge is passed on to tourists which deters some from making a visit. In effect demand falls from Q_P to Q_S. Thus the charge or tax not only raises revenue which can be used to meet the cost of environmental protection but also reduces the pressure of demand.

Grants and subsidies

Long-term environmental improvements can be encouraged by incentives in the form of grants to encourage investment in more efficient and up-to-date technology to reduce materials and energy use and the generation of waste, or even to eliminate some activities entirely. Subsidies are essentially a nega- tive tax and a recurrent form of investment grant. The problem with the use of grants and subsidies is that they can be extremely costly for governments if open-ended. Moreover, there is no certainty that there will be a sufficiently large take-up where schemes offer less than the total cost. Such intervention distorts the cost structure of particular industries if the take-up is uneven. It can also create international differentials, giving a competitive advantage to industries in countries where such instruments are applied in comparison with others in which they are not.

Economists' attitudes to subsidies are somewhat ambivalent because they have been conditioned by being concerned with identifying and evaluating the unintended outcomes of encouraging new entrants and existing, often inefficient producers, for example farmers, to remain in business giving rise to over-production, over-use of resources and environmental degradation. Nevertheless, there are circumstances, particularly with respect to situations involving public goods and externalities, where grants and subsidies could be beneficial. For example, in an urban context, conservation of heritage buildings is supported by public sector capital grants and subsidies or tax breaks because it is recognized that the maintenance costs of such artefacts to occupiers and owners is much higher than for modern properties. In this sense the level of grant, subsidy or tax break is an implicit recognition of the social benefits generated by the built and natural environment. Another important example is the change in agricultural policy in the European Union since the mid-1980s, whereby grants and subsidies are paid to farmers who farm in more traditional ways, thus enhancing the rural environment for wildlife and, where feasible, visitors.

Charges

The distinction between charges and taxes is not always clear and they are occasionally considered together, suggesting that they have similar charac- teristics. Charges can be applied to emissions of pollutants, supply of prod- ucts or services and consumers' use of environmental facilities and so are relevant to both the supply and demand side of tourism. Deposit and refund schemes might be considered as a variant of charges instruments, but increasingly they are being perceived as separate because they concern the conservation of material resources through recycling as opposed to environmental quality. For example, the hospitality sector purchases many commodities which are supplied in packaging and/or containers. To encourage and fund recycling of these, a number of countries have initiated deposit-refund schemes for metal and plastic beverage containers (Turner *et al.*, 1994). Charges can raise prices to consumers and almost certainly increase producers' costs and to this extent have the same effect as taxes.

However, there is a technical difference in that economists perceive charges as being levied to deter specific activities or the use of environmentally harmful products or to meet the full costs of output and/or services and facilities. Taxes, on the other hand, are conceived as meeting identified levels of consumption or production, especially in relation to maximizing net social benefits or (minimizing net social costs) by attaining an economic optimum (see Figure 8.1). The distinction can best be explained by illustrating briefly the various forms of charges.

Emission and effluent charges can be levied to deter the generation of particular wastes and can be effective if they are higher than the cost of the required investment, where possible, to eliminate discharges. They can also be employed to cover the cost of disposing of waste or the necessary remedial action to offset the effects of the emission. User and waste disposal charges are similar to emission charges where specific facilities or services are publicly provided to collect, treat, dispose of or recover waste materials. These charges should cover the net cost of providing the facilities or services. It is possible that the providing body can secure income from recycling some materials and therefore offset the gross cost. Commercial organizations are normally charged, which can result in illegal fly-tipping to avoid such costs and thus exacerbate environmental problems and render the charges counter-effective. There are some fears that this may increase in the UK where a landfill tax has been introduced which may raise the charges levied by local authorities. To this extent other charges which do not encourage such adverse behaviour are preferred.

Product charges are levied in order to deter the demand for or supply of goods and services which have a detrimental environmental impact. The charge could be set at a level which effectively makes the price of the product prohibitive so that demand and supply are entirely curtailed or at a lower level which decreases demand or supply. It may also induce substitution to more benign materials or services. The use of product charges is less effective where demand is inelastic or the volume of sales is small. Such charges can be applied at the output stage if they lead producers to modify their operations to make use of certain materials, or at the point of sale to reduce consumption. In both cases, a charge is essentially a selective indirect tax but, unless related directly to the environmental costs generated by the product, is not designed to affect consumption or production to achieve a social optimum.

Taxes

Levying a tax is the foremost economic instrument for achieving environmental improvements. Unlike charges it should, ideally, be set at a level which takes account of the social costs generated rather than solely the private costs of the activity. If correctly estimated, it induces a move to an economic optimal position which maximizes net social benefits and reduces environmental degradation. In theory, taxes can be applied on capital, for example, to deter the use of inefficient and polluting technologies, and/or products or activities which pollute.

Much debate in the economics literature is concerned with the merits of taxes (see for example Turner, 1988) in comparison with regulation. The principal argument is that with regulation, producers have no incentive to reduce pollution below the standard imposed which, in any case, may be arbitrary and bear no relation to the costs falling on society. However, notwithstanding other advantages of taxes, such as the lower cost of implementation, difficulty in avoiding them, the possibilities of inducing investment in new, cleaner technologies or substitution to less polluting products or processes, there are some disadvantages. These are mostly concerned with the practical difficulties of applying them, in particular the problem of accurately measuring the environmental costs on the one hand and the social benefits generated by production of the good or service on the other. There are even wider issues, for example, regarding the eventual tax burden and acceptability, but since these are common to any form of intervention which affects costs and prices, observations on them are deferred at this point.

Tax-subsidy devices

The use of a tax and subsidy approach to dealing with environmental problems is a relatively recent phenomenon in environmental economics and is exemplified by the deposit-refund, recycling credit and disposal schemes designed to facilitate the reclamation of materials. Their effectiveness depends on the relative costs of primary and secondary sources and the ability to develop the necessary structures to run the schemes which connect consumers with manufacturers through collection operators. A number of countries in Europe and states in Canada and the USA promote recycling schemes, most using a levy at either the production stage, which may be passed on to the consumer, or a charge on the product at the point of sale. Refunds are made as the materials are reclaimed either at public sector recycling locations or by suppliers. There is little evidence to indicate how efficient such schemes are and, unless supported by regulations restricting disposal of recyclable materials, they are unlikely to be adopted by manufacturers.

Quasi-price instruments: marketable licences, permits or quotas

Setting quotas as a means of controlling output or the use of a resource gives some indication of how markets are likely to react and operate. For example, there has been a long experience of production quotas in agriculture and international trade. More contentious are quotas relating to open and common access resources to allow for their replenishment, as in the case of fishing and whaling. Control of output and yield can have some beneficial environmental effects so it is but a short step to control directly the residuals of production, particularly pollutants, by granting licences or permits or quotas to emit them. Making them tradable gives more flexibility and allows the market to determine their value and producers to assess the best course of action to comply with a set standard.

The outcome is much the same as with other pricing mechanisms. If a particular producer finds that the cost of purchasing a licence or permit is lower than undertaking investment to reduce emissions, then a licence/permit will be chosen. Conversely, a business which finds the cost of abatement lower than that of the permit will sell any permit it might have been granted. It is of course conceivable, depending on the basis on which the initial allocation was made, for firms to sell a part rather than all of a pollution quota. For example, if the permit allows the emission of 100 tons of residual and the actual amount is only 80 tons, then the remaining tonnage can be traded.

The advantage of the permit/quota approach is that it potentially lowers the cost of compliance, can reflect the variations in cost structure of businesses and the dynamics, in terms of entry into and departure from an industry, of the market. The drawback of the instrument is the setting of the aggregate permitted residual, which raises the fundamental issue of the ability of the environment to assimilate it. Thus regulation ultimately has to be exercised in the light of experience, raising or lowering the permitted level when appropriate and monitoring whether firms comply. This may impose costs on society at large. There is some evidence of how the instrument works with respect to air emissions in the USA (Tietenberg, 1988). There are moves in other countries to extend the idea to the use of resources, along the lines of those already implemented in agriculture and fishing; for example, in the UK the idea has been mooted for access to the countryside, especially to environmentally sensitive areas.

Regulations: standards and targets

This form of instrument is of interest in economics because of the effect it has on consumer prices and production costs. It has also been widely applied in many countries with respect to air emissions, effluent discharges and noise, so that it is possible to assess its effectiveness in achieving environmental goals. In economic terms the monetary assessment of the cost of standards and targets in comparison with the benefits of achieving a reduction in pollution and/or an environmental improvement can be considered at two levels. The attainment of maximum economic efficiency requires that regulation should be judged by whether net social benefits are maximized. At a lower level it can be appraised on a cost effectiveness basis, the criterion being that benefits should exceed costs for a given environmental objective.

The appeal of regulation is that it may obviate the need to estimate the required level of, for example, pollution or resource use, its certainty in attaining specified targets, its transparency, its acceptance by the main sectors of society, especially industry, its relative simplicity in monitoring and enforcing compliance and the placing of responsibility for meeting identified standards on those causing problems. On the other hand, in most countries there is no coherent regulatory strategy. Most policies have been introduced to deal with particular problems and promotion of national, let alone international, standards and regulations, is still in its infancy. For example, the environmental mediums of air, water and land have been considered individually and it is only recently that a more integrated approach has been advocated, for instance the adoption of best practicable environmental option (BPEO) in the European Union. These standards tend to be technologically based, in which the cost of the investment required to meet them is taken into account, i.e. should not be excessive. Nevertheless, with increasingly rigorous standards and consumer pressure to effect environmental improvements, the burden on producers is likely to increase, with cost consequences which may be passed on in the form of higher product prices. Moreover, greater inflexibility may be introduced the more aggregative the regulation so that variations in local and industry level conditions are not reflected. It is possible that the effectiveness of regulation may diminish over time because those subject to it will influence the body responsible for administering it. This occurs because both parties need to confer to implement regulations. An analogous example is the procurement of planning permission by property developers which has been termed 'rent seeking' (A.W. Evans, 1982), particularly when there are opportunities for mutual benefits to both parties, for example 'planning gain' whereby the community secures, say, leisure facilities within an allowed development.

SOME GENERAL OBSERVATIONS ON POLICY INSTRUMENTS

Although it has not been exhaustive, this review of policy instruments and their respective advantages and disadvantages has highlighted some of the more important issues arising from the means for achieving environmental goals. Some common economic evaluation threads are discernible in the examination of the instruments, whether priced or regulatory. A key factor from an economic viewpoint is whether the measures are cost-effective and/or efficient. These criteria will not be rehearsed at this point except to state that although conceptually it is possible to appraise the effect of the instruments, in practice it is extremely difficult, especially those which are regulatory. In general economists conclude that price instruments are both more cost-effective and efficient.

A second and crucial issue is that for nearly all the instruments considered, certainly for taxes, charges and tradable devices, there is an implicit assumption of an identifiable limit to the marginal social cost (or benefit where grants and subsidies are being offered) which the instruments will prevent being exceeded. Two observations are apposite. The first is the practical difficulty of ascertaining the allowable environmental damage which is the foundation of the concept of limits to acceptable change as an environmental target. The second is that the economic optimum towards which the instruments are meant to guide activity does not necessarily accord with optimal positions propounded by other disciplines, such as ecology. For example, the European Union Bathing Waters Directive identified standards (environmental indices) which are relevant to human health and consumer satisfaction but which are not relevant to marine biodiversity. Regulation may be more arbitrary if it is decided that a reduction in environmental damage is desirable *per se* without any reference to an economic or any other optimum.

Another aspect which emerges is the equity issue, in particular whether the polluter pays principle is actually effective. A further dimension of this issue is the degree of regressiveness (i.e. lower income groups pay proportionately more than higher income groups) of the instruments, irrespective of whether the burden falls on consumers rather than producers. The basic economic concept of elasticity of demand and supply gives insights as to the possible burden while knowledge of variations in demand elasticity over different incomes offers evidence on the degree of regressiveness of specific instruments. Clearly the more inelastic demand is for products in relation to supply, the more the burden of an environmental charge or tax falls on the consumer. Priced instruments, especially if they are specific rather than *ad valorem*, i.e. levied at a flat rate rather than on a percentage basis, in essence being in their effect like indirect taxes, are usually regressive. Their impact can be offset by governments compensating lower-income groups, but using revenue from charges and taxes in this way may reduce the funds available for environmental improvements where the public sector is responsible for doing so. Much the same arguments apply to the differential effect on specific firms or industries where the instruments distort the allocation of resources.

Passing reference has been made to the implementation of the instruments but this aspect needs to be spelt out more fully. There are a number of evaluative factors for judging whether instruments are both effective and workable. Effectiveness concerns the achievement of particular aims or objectives, whereas the extent to which instruments are workable is more a function of cost and feasibility. Costs arise when there is a requirement to administer an instrument. It becomes necessary to ask whether such administration can be undertaken through existing structures, such as the fiscal authorities for the collection of charges, taxes and payment of subsidies or tax relief, the land-use planning system for regulations and specialist environmental bodies for licences, or permits or quotas, as opposed to setting up specific organizations and structures. There may also be a requirement for monitoring activities and enforcing compliance. In addition there are the operating costs incurred by those subject to instruments, especially the ones more firmly set in a market context, for example, the transaction costs of securing a tradable permit or quota. Other factors determining whether particular instruments are feasible or not are the ease or difficulty with which the nature, sources and possible sufferers of environmental problems can be identified, the possibilities of ascertaining and measuring the extent to which their impact has unintended and detrimental effects and how far they are considered acceptable by society in general and the business sector in particular.

ENVIRONMENTAL POLICY INSTRUMENTS IN THE CONTEXT OF TOURISM

Tourism, like other economic activities, is a consumer of substantial quantities of materials and energy and generator of wastes. For example, the impact on the stratosphere of burnt aircraft fuel is a significant polluter. Tourism activity also tends to be concentrated in specific areas, often over-exploiting scarce resources such as water and creating waste disposal problems and pollution. Accordingly, many tourism activities should be subject to environmental policies and instruments that are applied to productive industries. Yet tourism possesses characteristics which mark it off as different and therefore, potentially, requiring a different means of dealing with environmental problems. First, the generally fragmented nature of tourism supply results in the identification of the source of its environmental effects being more difficult, while periodicity and seasonality give rise to an uneven temporal impact (Sutcliffe and Sinclair, 1980). Second, that impact tends to be physical in terms of sheer numbers of people (therefore exceeding the capacity of transport modes and tourism resources); visually intrusive built environment and evidence of tourists (such as cars parked in scenic areas); over-use of natural environments leading to degradation, disturbance to wildlife and reduced biodiversity; waste generation. Third, perhaps reflecting an implied criticism of mainstream environmental economics, tourism, in common with a number of other activities, has the potential to generate considerable benefits. This aspect tends to be submerged in the economic analysis of policy instruments which dwells almost exclusively on costs.

With respect to combating detrimental environmental effects concerning the use of materials and energy, the instruments examined are clearly relevant in so far as service activities, such as tourism, consume substantial quantities of natural resources-based inputs. Certain policies are virtually specific to manufacturing, where there are tangible products and consequent chemical effluents or emissions are generated, so that the blanket imposition of

207

emission and product charges, permits and quotas would, prima facie, not appear to be applicable. This would suggest that taxes and subsidies, deposit and refunds and regulation seem to be the most appropriate instruments for dealing with the environmental problems generated by tourism, reflecting its specific characteristics identified above. The issue of jet aircraft emissions, an instance of an emission whose source is obviously identifiable, has not been solved by any of the currently advocated instruments. The growth of air travel is accelerating, particularly in the long-haul market segment which includes a significant proportion of eco-tourism. Thus the contribution of tourism to global environmental problems may be increasing in relation to that by manufacturing industries, some of which have a wider choice in adopting innovative and less polluting technologies. It is extremely difficult to combat this problem of jet aircraft emissions into the stratosphere. While in the short run draconian environmental charges, by substantially raising the price of air travel, might deter tourism travel, it is unlikely that the number of flights would be drastically reduced; aircraft would merely fly with lower pay-loads which is both technically and economically inefficient. The solution is more likely to be a long-term one, being dependent on technological developments which reduce emissions or involve the use of less polluting fuels, or even a new means of propulsion.

Tourism concentration in certain locations suggests that instruments designed to deal with local problems, such as waste, litter, noise, degradation of physical resources, visual intrusion and cultural and social deterioration, are particularly suitable. In a sense the hospitality activities involved in tourism are much the same as those of households but on a larger scale, with some spatial concentration, such as in seaside or skiing resorts or certain districts in historic urban areas. In this context, the issue is largely one of materials and energy use and waste disposal, requiring instruments aimed at encouraging reductions in the use of packaged materials or energy saving. User charges to meet the full cost of waste disposal, combined with product charges levied on manufacturers of products used, perhaps related to the amount of packaging, would provide an incentive to cut consumption. Similarly, deposit and refund schemes would encourage recycling, where feasible. Reductions in energy use can be stimulated by capital grants to use more efficient heating and lighting systems and to improve insulation. Materials and energy use and waste generation are the aspects of environment issues which have been targeted by firms themselves and, as shown, the potential commercial pay-offs determine the likelihood of the initiatives being successful. The dissemination of information, which demonstrates the advantages of implementing improved systems and management, is also relevant. The impetus for buttressing such market actions with publicly applied instruments is often local, suggesting administration at the local government level.

It is, however, the second characteristic of tourism which is more crucial and intractable. The physical impact of tourists and the provision of their needs threaten to destroy the resource base, particularly where natural and fragile environments are involved. This pressure is aggravated by the periodicity and seasonality characteristic of much tourism. Some environments are often at their most vulnerable, for instance in bird and animal breeding seasons, when visitor numbers are highest. In other cases, attempts to reduce numbers at weekends or in the high season by encouraging off-peak and off-season tourism are not efficacious; sustained use over the year can be worse for some resources, for instance hiking can erode paths more severely in wet and/or winter periods.

The problem of concentrations of large numbers of tourists suggests that controls are required in some destinations (Wanhill, 1980). One obvious approach to the problem, in addition to the use of the price mechanism, is to use regulatory instruments on both the demand and supply side to restrict visits. Some reference has already been made to means of restricting access in earlier discussions of problems concerning renewable resources. The land-use planning system can act on the supply side by restricting the accommodation, facilities and services made available. This, however, would be very difficult to implement for existing destinations since it might involve demolishing a proportion of the built environment. Another approach is to make access more difficult by quotas or controls, for example the number of flights, by issuing only a limited number of visitor licences to tour operators or permits (akin to a visa) to tourists. Examples of this kind of control already exist; entry into the Bahamas, national parks in North America and the 'take' in hunting or fishing are cases in point. Whether these licences or permits would be tradable might depend on the circumstances. A more firmly market-based method would be to levy taxes on suppliers who exceed set limits, or entry taxes, perhaps at a punitive rate, to control numbers. Charges and taxes are already used but not at a level that is likely to deter entry. Indeed, the fairly widespread operation of an airport tax on passengers is an example. This form of tax is used as a revenue-raising device but is unlikely to be applied to mitigate the environmental impact of tourism. There are exceptions; for instance, in Vancouver the exit tax on air passengers is specifically stated as being for the improvement of the airport and the environment adjacent to it.

The problem with physical or financial means of controlling numbers is that inefficiency and inequity are likely to result. Economies of scale in providing travel, accommodation, facilities and services may be lost on the supply side, while if increased costs are generated and passed on to tourists, or a direct charge on them made, those in lower-income groups may be priced out of the market. Moreover, those working in tourism sectors in host communities may be disadvantaged if the level of activity is reduced. There are also other factors which may create significant costs, for example the administration of the instruments or simply whether they are cost effective. Other issues raised are their compatibility with environmental and regulatory objectives, their acceptability and what kind of incentives they offer to tourists and firms to find their own solutions which, on the one hand, might circumvent the restrictions but, on the other, obviate the need for publicly applied instruments. Rather than attempt to restrict numbers, an alternative approach, which might resolve the problem of overcrowding in existing destinations, would be to levy user charges and allocate the revenue raised to institute environmental protection or reparation. This strategy is not favoured by many environmental economists for it smacks of attempting to cure the affliction as opposed to preventing it from occurring.

Thus reducing the use of materials and energy and the generation of waste and controlling numbers of tourists can involve the application of virtually all the policy instruments examined and evaluated. Current policy thinking has not advanced so far as to perceive the operation of service activities in the same light as manufacturing activity. Consequently it has not considered the full range of instruments as being applicable as indicated here. Given the potential and actual detrimental effects of tourism, the time is approaching when restrictions on its development and operation almost certainly have to be more severe.

To this point, the evaluation of policy instruments in the tourism context has echoed that given in the literature of emphasizing the environmental costs of economic activity. However, the many possible beneficial effects of tourism mean that the instruments presently proposed might counteract such effects. The general consensus in economics is that beneficial externalities should be enhanced, the means being to compensate or subsidize those making such provision whether they be in the private or public sector. Prime examples are the occupiers or owners of listed heritage buildings or sites who incur additional costs to conserve them. Both grants for capital costs and subsidies for recurrent expenditure are seen as the principal instruments. However, many tourism resources are natural and open access environments so that property rights do not exist or cannot be exercised. With respect to unique resources of global significance, such as Antarctica, the oceans, rare flora and fauna, whose importance transcends national boundaries, the possibilities of allotting grants or subsidies are, at the very least, problematic. It is necessary to establish international bodies and agreements to oversee and manage such resources. This means instituting international regulations and administrative structures.

Past experience with conventions on whaling and fishing give rise to pessimism that agreements can be made binding and adequate funding made available for enhancement and/or policing of resources, although the position in Antarctica gives cause for a greater degree of optimism. It appears that some kind of ownership and thus property rights, enforced through international law, need to be introduced. These kinds of issues go beyond mere implementation of policy instruments which are applied at a specific and operational level. Overall, it would appear that policy instruments advocated in environmental economics is equally applicable to service activities such as tourism. However, their application needs to be assessed within the wider context of the global environment and strategic policy-making if the instruments are to be effective in both a restrictive and positive way.

CONCLUSIONS

In this chapter the review of environmental analysis, in economics in general and tourism in particular, prompts three main observations. The first is that conceptual and theoretical developments in economics have not been fully incorporated into the analysis of tourism. The second is that there are some disparities between the policy desiderata proposed in economics to achieve environmental goals and those recognized as important in tourism. The third is the continuing debate in economics on the extent to which market intervention is necessary to secure sustainable development, which spills over to create uncertainty of commitment in practice.

There is, thus, a need for economists to ensure that the substantive prescriptions of environmental economics and its policy implications are brought to the attention of those involved in applied work. Nevertheless, it would be improper to suggest that environmental economics is a fully mature and comprehensive system of thought and that its methods have been sufficiently developed to be applied in all contexts. Some elements of the subject are already well advanced and applicable. For example, the concept of total economic value, the CVM, HPM and TCM methods for ascertaining the level of non-priced demand for such objectives as environmental conservation and the policy instruments for attaining it, are well developed and operational. On the other hand, the analytical framework in which to appraise environmental policies is still being debated in economics. This is certainly so with respect to the boundaries of cost-benefit analysis in terms of what benefits and costs to include and their magnitude; in short, when are the impacts so minuscule as not to warrant measurement? A further subject of debate is the rationale for and means of taking account of intra- and inter-generational distributional effects.

A more fundamental topic of controversy is the extent to which it is necessary to intervene. The market failure concept, as suggested earlier, is not accepted universally in economics. There are those (for example, R. Coase, 1960; Demsetz, 1969; A. Randall, 1993) who suggest that the existence of such factors as externalities and public goods are merely imperfections which can be resolved by the market if property rights are properly defined and internalization can occur.

Yet another area where environmental economics is still in the process of being developed and made operational is where its concepts and analysis rest on a scientific base. For example, the rate of consumption which allows renewable resources to replenish themselves and the establishment of the

211

levels of emission or effluent discharge which enable the environment to assimilate pollution is a scientific issue. Thus, the concept of maximum sustainable yield, notwithstanding the economic limits of exploitation being determined by benefits and costs, is founded on the physical/biological imperative of the ecological processes involved. Hence the arguments put forward within environmental economics must be made in conjunction with those emanating from the natural sciences (Hohl and Tisdell, 1993).

The issue of environmental protection and sustainability is also political. Leaving matters to market forces has been called in question. Some environmental economists assert that the discipline should be more persuasive in showing the environmental implications of the current emphasis on economic growth and, in turn, tourism expansion (D.W. Pearce et al., 1989; Cater and Goodall, 1992; Goodall and Stabler, 1994). The organization Redefining Progress (1995) reflects these concerns in advocating, as a start, that national income measurement should account for the impact of economic activity on the environment. The evidence of global warming, ozone layer depletion, acidification, deforestation, desertification and de-biodiversification is cited as pointing to the need to move from the implicitly accepted weak, or at best intermediate, sustainable development positions, to the strong one (Turner et al., 1994). The weak stance allows for the substitution of human-made for natural capital and the reliance on innovation via the market whereas an intermediate position, although taking account of the constraints imposed by not allowing renewable inputs to be degraded, still considers that technical progress will resolve problems. The strong position argues for the non-substitution of natural capital for that which is human-made but goes further in advocating the precautionary and safe minimum standards principles (D.W. Pearce et al., 1989) where uncertainty prevails, particularly when decisions on resource use might introduce irreversible trends. It may imply lower economic growth and possibly no enlargement in the scale of activity.

With respect to tourism, reflecting these prescriptions, there is a requirement to resolve the extent to which sustainable tourism is consistent with maximizing its present value over time, i.e. maintaining its future commercial viability. In addition, the conventional economic analysis concerning market failure, which advocates an optimal position where marginal social costs equal marginal social benefits, may be incompatible with an ecological optimum and so lead to the long-term degradation of the environment. Furthermore, ignoring the differential spatial impact of any environmental policies applied may increase materials and energy use; for instance, some areas may have a greater capacity to absorb pollution than others and therefore the imposition of common policy instruments may be unnecessary and wasteful. A more selective approach is required. Sustainable development and tourism concepts and policies have originated in industrialized economies where there is already a high proportion of human-made capital.

The pursuit of sustainability is not only more likely but has a greater chance of being achieved in these countries. To impose the same regime on developing countries whose environmental problems may be very different smacks not only of a measure of arrogance but also may be inappropriate (Curry and Morvaridi, 1992).

Research, both in general and specifically in relation to tourism, which needs to be undertaken to move towards sustainability can be summarized as:

- the systematic investigation of the scale and nature of the environmental impact of tourism activity at the local, regional, national and global scale
- the establishment of a real understanding of consumers' preferences for environmental protection and sustainability rather than the views represented by tourism firms for commercial gain
- the examination of the ecological and economic concepts of and prescriptions for sustainability and the means of reconciling them, recognizing that the holistic and stochastic interrelationship of ecosystems is unlikely to yield deterministic perspectives and solutions
- the development of environmental policies based on the notion of safe minimum standards.

At present the focus has been limited to defining sustainable development, considering how best to measure it, together with its weak, intermediate and strong scenarios and their policy implications. A major obstacle to progress in conserving the environment continues to be the establishment of its economic value. It is crucial that environmentalists convince governments and businesses that the environment is not a disposable commodity but an integral part of the resource base, as are materials and different forms of energy. Moreover, its conservation is not always an additional cost but can, in fact, be the reverse.

Attitudes in tourism firms reflect what can only be described as the complacent political stance. Currently lip-service is being paid to sustainable tourism and environmental conservation. The underlying philosophy remains that any environmental action should not undermine the viability of businesses. The response to requests for environmental responsibility has been piecemeal and muted. Incentives to pursue environmental goals more vigorously will need to be offered to encourage tourism firms to safeguard their resource base and reduce the generation of detrimental externalities. They also need to be convinced that some regulatory measures may be needed. Economics, through its tenets and analytical approach, can demonstrate both the costs and benefits of environmental action and where these fall. It is a social and political, and perhaps even ethical, matter to decide on the distribution of such costs and benefits in society.

9

CONCLUSIONS

The principal aims of this book have been to demonstrate the ways in which economic theories and methods can help to explain a wide range of issues in tourism, to provide a more explicit and advanced theoretical basis for the critical evaluation of tourism studies and to indicate the implications for their empirical investigation. Economic methodology has the ability to explain, predict and quantify tourism phenomena, as was illustrated in the previous chapters. The approach which has been used in the book to demonstrate the contribution which economic analysis can make in these areas was to start by considering tourism as though it were a simple, undifferentiated commodity supplied, without constraints or externalities, in an unspecified set of markets and then to introduce progressively further concepts into the analysis in order to provide a detailed explanation of the complex nature of tourism. Thus, initially tracing the literature on tourism demand, the discussion commenced by treating tourism as an aggregate product, considering the individual and then the social nature of tourism purchase decision-making. The subsequent examination of tourism supply first identified the different components of tourism products and the characteristics of the markets in which they are provided, before developing the analysis to examine the dynamic nature of markets and the strategic nature of the interrelationships of the firms operating within them. The advantages of examining tourism demand and supply within an international context were then demonstrated. Finally, the importance of the environment as an element of tourism supply, the issues of market failure and externalities and the possible case for related policy intervention were considered, reflecting current advances in environmental analysis. By proceeding in this way from the simple to the more complex, it was possible to introduce a large number of concepts and analytical approaches which can be used in tourism analysis. It is not possible for the coverage to be completely comprehensive and there are consequently some omissions, such as spatial theory and labour market theory. The lack of consideration of these areas in no way decries the important contributions to the study of tourism which have been based on them, for example by urban economists and geographers. Reinforcing the theme pursued throughout the

214

book, the aim of this concluding chapter is to identify the main contributions of economics to an improved understanding of tourism which have been made in the preceding chapters and to highlight some of the many issues requiring further research.

The microfoundations of the demand for tourism have been somewhat neglected in the existing tourism literature; therefore the main contribution of Chapter 2 was to provide a theoretical analysis of them. The relationship between holiday purchases, the expenditure budget and the potential tourist's preferences was examined and the concepts of substitutability or complementarity in tourism demand were introduced, as well as the effects of income and price changes on demand. These concepts are particularly useful in tourism marketing, indicating which tourism products or destination countries are strong competitors, and with which joint marketing efforts might be undertaken. The theoretical discussion of the microfoundations serves as a vital basis for the formulation of models for explaining and predicting tourism demand accurately and for evaluating the studies undertaken to date. Chapter 3 used the preceding analysis along with further insights from mainstream research in a critical discussion of the two main approaches which have been used to estimate tourism demand. The discussion questioned the theoretical basis, and hence the empirical results, of many of the studies which have used the first approach, single equation models. In contrast, it was shown that the second approach, system of equations models, provides a useful basis for future research in so far as the microfoundations of demand are incorporated appropriately when aggregating from individual to total purchases of tourism. However, as in single equation models, tourism has been treated as a commodity which is generally undifferentiated other than by destination and which is purchased by consumers who are unspecified other than by nationality.

The analysis of demand in Chapters 2 and 3 raises two related issues. The first is the social context of decision-making, which is not taken into account by the individual-based economic theory of consumer demand. For example, the theoretical discussion in the first part of Chapter 2 considered an individual's holiday decisions in isolation from the person's social reference groups. It might be argued that individuals' preferences are revealed by their behaviour with respect to tourism purchases (the concept of revealed preference). However, this does not acknowledge that people's notional desires may be constrained not only by their income but also by social factors such as peer pressures or intra-household power relationships. Decisions are also related to such variables as the information available from family, friends and acquaintances. Thus, the microfoundations of demand should incorporate the social basis of decision-making. This is a fundamental area for investigation by mainstream economists in which contributions from other disciplines may be enlightening. The tourism research agenda could include examination of the process of the formation of preferences for tourism relative to paid work, unpaid time and other goods and services by individuals, within both the individual psychological context governing motivation and the wider one of the family or social reference group, and their translation into tourism purchases. The dynamics of changes in motivation, preferences and expenditure over time also require investigation.

The second issue is the relationship between the demand for tourism as a composite commodity and the demand for the different components of which it consists. The review of the models which have been used to estimate tourism demand, in Chapter 3, showed how analysis has focused on tourism as an aggregate commodity, disregarding the fact that consumers actually purchase holidays with different mixes of characteristics. Consumers' purchases of particular types of holiday involve derived demand for sets of holiday components, such as specific types of transport and accommodation in desired locations. Hence, it is necessary to examine the demand for each of the components in order to explain the demand for tourism as a whole. This type of approach is exemplified by the early work of Lancaster (1966), who considered the demand for goods in terms of the demand for sets of characteristics, and has been applied in such forms as hedonic pricing models. Analysis of the demand for sets of holiday components would help to identify the relationships of complementarity or substitutability between different components, by different individuals or groups of consumers. The approach would also be consistent with the examination of the supply of tourism components by a set of interrelated industries and markets, undertaken in Chapters 4 and 5.

Overall, it appears that the focus of tourism demand research should be on its investigation at more disaggregated levels, in terms of both its microfoundations and the derived demand for individual components of the tourism product, as a means of providing an improved explanation of demand at more aggregate levels. Investigation at both disaggregate and aggregate levels should increase the accuracy of not only the results obtained from the demand models themselves but also those generated by forecasting models which are based on them. Underlying all tourism demand models is, of course, the distribution of income and wealth in the economy. Studies of demand at the microeconomic level of individuals and social groups could provide some indication of the ways in which the level and pattern of demand for tourism, as well as for other goods and services, might change following an alteration in distribution, for example, a change in the tax and benefits structure.

The economic analysis of tourism supply undertaken in Chapters 4 and 5 was innovatory in providing a range of analytical approaches which can be used to examine tourism firms and markets. Initially, neoclassical theory was used to examine the operation of firms – price-setting, quantities supplied and associated levels of profits – in market structures ranging from monopoly to perfect competition. Application of the theory to different tourism components (accommodation, transport and intermediaries) demonstrated the advantages of a disaggregated analytical approach by showing that different market structures apply to different tourism components. Moreover, within a given country, the markets for particular components may exhibit more than one type of structure, as in the case of UK tour operators offering package holidays, where the large firms operate under oligopolistic conditions and the smaller operators exhibit monopolistically competitive features. This raises the issue, that appeared in the context of tourism demand, of the level of aggregation at which to undertake supply analysis. In the case of tourism supply it appears that, for some components, the analysis should not assume that a single market structure prevails but instead identify and examine the set of market structures which are relevant. Hence, further examination of the structure and operation of the markets supplying each component of tourism, along with their interrelationships, is necessary.

The application of the theoretical models of market structure to key components of tourism supply revealed a number of features which should be considered more extensively and rigorously in empirical studies of tourism in different countries and contexts. Some are indicators of market structure, such as the number and size of firms, the degree of market concentration and level of entry/exit barriers, economies and diseconomies of scale, capital indivisibilities, fixed capacity and fixed costs of operations, which have been insufficiently examined in past studies of tourism. Other features also refer to the strategies which are pursued by firms in imperfectly competitive markets and are price discrimination and product differentiation, and pricing policies – leadership, wars and market share strategies. Examination of these features in specific contexts would aid the identification of the type of market structure which most nearly applies to a particular tourism sector or sub-sector. For example, evidence of imperfectly competitive structures, particularly monopoly, would indicate a case for further investigation of the operation of firms in the market and the possible need for policy intervention to limit the prices charged to consumers and the associated level of profits which firms make. Since the strategies which firms pursue may have the effect of changing the prevailing market structure, research on the interrelationships between market structure, firms' conduct and their performance is also necessary on a more holistic basis.

The SCP paradigm, developed within industrial economics, is a broad approach which can describe the interrelationships between structure, conduct (operation) and performance. In Chapter 5 the application of the paradigm to UK tour operations showed that while the approach provides a useful framework for identifying and describing the key characteristics of structure, conduct and performance, it takes the market structure as given and cannot explain the dynamic nature of markets and firms' strategies within them. Thus, while it is useful as an expositional device, it has limited analytical value. In the same chapter it was shown, in a review of developments in industrial economics, that the scope of the analysis of supply has widened beyond its neoclassical foundations but that new contributions, such as evolutionary, institutional, principal–agent and transaction costs economics, are tending to be developed independently rather than being welded into a more cohesive theory. Likewise, economic psychology and behavioural approaches which address supply issues are somewhat set apart. Few of these recent developments have crept into the tourism literature. This is unsurprising given the current state of the art within economics and the discipline's neglect of tourism. However, even at these relatively early stages in the derivation of the new methods of explaining and predicting supply, economics can contribute to improving the analysis of tourism.

The core feature of Chapter 5 is the proposition that in game theory lies the solution to developing a feasible theory of interactions between firms. Although it requires refinement within economics, game theory is clearly more appropriate than the traditional comparative statics neoclassical framework for examining the dynamic interrelationships between firms and the different strategies which they might adopt concerning pricing, the level of output produced and product differentiation, as well as the possible effects on market structure and their own performance. It can incorporate insights from other schools of thought and has the advantage of identifying explicitly the range of feasible strategies and outcomes. Therefore, the great challenge for both economic and tourism research is to apply game theory in empirical studies of supply in service as well as manufacturing sectors.

The analysis undertaken in Chapter 6 was hampered by the fact that mainstream international economics has neglected many service activities, notably tourism. Similarly, tourism literature has taken little account of the issues examined within international economics. Therefore, in the first part of the chapter, a new attempt was made to indicate the ways in which concepts and theories developed in international economics may provide insights into tourism. However, the discussion also demonstrated the limitations of traditional trade analysis, set within an assumed context of perfect competition, by showing that although theories of comparative advantage can provide a broad explanation of cross-country patterns of production and trade, theories which take account of imperfect competition at the international level are relevant. These theories encompass such features as the demand by tourists for holidays of different qualities, the role of economies of scale and scope, product differentiation and market segmentation, competition via research, development, innovation and imitation, the international interdependence of firms and the associated effects on market structure at the international level. The discussion of economic integration between firms is particularly important and a relatively neglected area within the tourism literature. It furthers the analysis of tourism supply by showing that firms can use alternative types of integration as strategies for altering market structure and, hence, prices, output and profitability in an open economy context. Examination of the nature, causes and effects of cross-border integration between firms supplying different components of international tourism is also important because of the significant welfare effects which alternative types of integration can bring about.

The main implication of the open economy analysis in the first part of Chapter 6 is the need to extend and deepen the analysis of the demand for and supply of different components of tourism by incorporating insights from international economic theory into empirical studies. Game theory analyses of strategic integration between tourism firms could be of particular relevance and have not, as yet, been undertaken. The prevalence of imperfect competition in tourism markets also suggests that there is scope for strategic tourism policies by governments which attempt to benefit firms of their own nationality. The characteristics of such policies, for example the imposition of regulations and modification of the institutional context for tourism, warrant investigation as they can alter the ground rules for competition in the global context.

Some key effects of international tourism were considered in the second part of Chapter 6 and it is useful to identify the related implications for research. These include further investigation of the sectoral, spatial and socio-economic distribution of the additional income resulting from tourist expenditure and the associated welfare repercussions. The nature of the demand for and supply of labour in different tourism sectors, for example, has been the subject of detailed attention for only a small number of countries and the effects of international tourism in changing existing employment structures has been examined for an even smaller number. Investigation of tourism employment is necessary and could incorporate both mainstream labour market theories and feminist approaches. Debates about the repercussions of instability in foreign currency earnings from tourism are related to mainstream debates between Keynesian, monetarist and neoclassical economists and one means of moving them forward would be to incorporate, into tourism studies, recent research on consumption, savings and investment functions in developing as well as developed countries. Applied studies could usefully focus on individual countries or groups of countries with similar characteristics. An interesting and comprehensive framework for explaining and estimating a range of effects of tourism, within an international context, is the computable general equilibrium (CGE) modelling approach which has only just entered the tourism literature and which clearly merits further attention.

New analysis of the relationship between tourism and economic growth, considered at the end of the chapter, raised the more general issue of the contribution of microeconomic sectors to growth at the macroeconomic level. This issue is related to major theoretical debates between mainstream economists. Keynesians view increases in demand, in the context of surplus capacity and the absence of a balance of payments constraint, as contributing towards higher growth whereas neoclassical economists traditionally regarded higher growth as arising from exogenous technological innovation. More recently, new growth theorists have made the case that growth may be determined endogenously (within the system). Keynesians see an obvious role for the government in boosting demand in the context of demand deficiency and the provision of incentives for increasing the demand for tourism within the economy would be an indirect means of doing so. Neoclassical economists, on the other hand, view the appropriate role for government as that of ensuring that markets work efficiently, implying that tourism development should be left to market forces. New growth theorists see a possible role for government in providing human or physical capital, including infrastructure such as roads or airports, in order to ensure that the marginal product of capital does not decrease and growth is maintained. However, they do not regard government intervention as necessarily beneficial. These are obviously fundamental debates which have influenced the formulation of economic policy at both national and international levels and the chapter provides the first discussion of their relevance to tourism and tourism policy-making.

It is clear that investigation of the types of tourism investment and policy, if any, which might contribute to economic growth is necessary. Research on the relationship between a country's comparative advantage in tourism or other economic sectors and economic growth would also be useful. Examination of the conditions under which a country can change its comparative advantage (dynamic comparative advantage) and move on to a higher growth path has barely begun within international economic theory and the role of tourism within the wider scenario has been ignored. As yet, there have been virtually no economic analyses of the causes and effects of specialization in different components of the tourism product or the role which the government might play in promoting a change in production patterns. Hence, a key implication of the discussion of tourism and economic growth is that issues in tourism should be considered in relation to core debates between economists about both the ways in which economies are conceptualized as operating, including the perfect or imperfect nature of markets, and the associated role for government policy-making. This implication is applicable not only to the specific area of tourism and economic growth but also more widely, as was demonstrated in the following chapters.

The environmental issues examined in Chapters 7 and 8 reflect current concerns both in economics and tourism. An attempt has been made to consider these in a broad way, acknowledging the far-reaching ramifications of all economic activity and the fundamental importance of the environment as a supplier of materials and energy life support system, and waste disposal sink, quite apart from acting as the tourism resource base. In this respect the environment acts as an input similar to other factors of production, such as human and physical capital, and is a crucial component of tourism supply.

In the early part of Chapter 7 it was posited that a number of areas of analysis hitherto treated as separate in economics, such as productive resources and energy use and conservation, resources valuation, environmental protection and sustainable development, should be considered as elements within the all-encompassing field of sustainability. Another, related matter which emerged was the necessity for an interdisciplinary approach to environmental issues, given that economic principles and policy measures with regard to, for example, the optimal use of resources over time, maximum sustainable yield and pollution abatement, depend on ecological, social, cultural and political factors. An imperative identified was the need for an analytical framework for appraising actions on the environment which accommodates use and non-use benefits, and the assessment of intra- and inter-generational distributional effects in addition to resource allocation effects. In traditional cost-benefit analysis, a basic project appraisal method in economics, intra-generational impacts are normally excluded as they carry the danger of double-counting and notions of the concept and measurement of total economic value have not been recognized. The identification of the necessity for a broader interdisciplinary approach within a more comprehensive analytical framework demonstrates the need for further methodological developments within economics in order for the discipline to make a more adequate contribution to attaining sustainable development. Nevertheless, although economics is still in the process of developing both its stance on environmental concerns and methodologies for its protection, it was shown in Chapter 7 that the subject aids understanding of more specific issues in tourism. For example, the concept of market failure, because of the existence of public or collective consumption goods and externalities, explains why tourism's very life-blood, the natural and human-made resource base, is under threat from growth in demand and the provision made to meet it. The open access nature and resulting degradation through over-use of this resource base is a key issue which has to be resolved in tourism as it does in other spheres of economic activity.

Whereas in Chapter 7 economic principles and their potential application to environmental problems were identified, in Chapter 8 discussion concentrated on the translation into action of statements of intent, as exemplified in the Rio declaration on sustainable development. This problem is central to the whole debate on environmental protection and therefore considerable attention was paid in the chapter to economic methods of valuing resources which are not traded in the market and the policy instruments which can be employed to secure environmental protection and, ultimately, sustainability. The justification for emphasizing these two aspects of environmental analysis is, first, that they, like those discussed in Chapter 7, raise issues in economics as well as tourism. Second, fairly extensive examination is necessary because in the tourism literature they are not considered in sufficient detail to indicate their relevance to solving tourism environmental problems.

The methods advocated for the valuation of use and non-use and existence benefits in economics are still being developed. While there is a broad consensus

221

on the conceptual acceptability, albeit with reservations, of the hedonic pricing, travel cost and contingent valuation methods, there are considerable technical difficulties in applying them. Moreover, empirical evidence, despite an increasing number of studies, is still patchy and somewhat inconclusive. Thus there is a considerable way to go before the methods are sufficiently refined and robust to assist the enactment and implementation of policy and to be acceptable politically. In relation to the evaluation of policy instruments, such matters are less fundamental simply because price-based or market instruments have already been tested and their effects monitored. There is, therefore, rather less need for further extensive research on the appropriate instruments, despite their limited application as means to control natural and human-made resource demand and supply with regard to tourism use.

What is surprising in considering environmental issues is that within tourism the impact of current trends is hardly recognized. There is a lack of awareness as to the implications of sustainable development tenets as a long-run aim and environmental protection objectives in the short run for future levels of tourism activity. In a sense this is demonstrated in publications on the subject which propose voluntary means of lessening its adverse effects. Environmental economics and its policy prescriptions have made little impression on either the tourism business sector or the tourism literature. This is partly a reflection of the lack of political will to implement more interventionist policies and partly that economics has failed to convey both the urgency and extent of remedial action required. Accordingly, the state of the art of environmental economics is less important than its inability to indicate the relevance of its principles and methods. This in itself is a problem which applied economists need to tackle.

Overall, this book has highlighted many areas of analysis which warrant further investigation in the context of tourism. The authors have attempted to demonstrate, by proceeding from the simple to the more complex, the nature of tourism as an atypical product and the economic concepts required to analyse it fully. While it is recognized that this text does not provide a complete analysis of the economics of tourism, it is hoped that it will constitute the point of departure for future research which will take account of the fact that tourists demand composite products, supplied by a collection of firms and industries which usually operate within imperfect domestic and international markets characterized by externalities, public goods and free riders. Tourism clearly tests economic concepts and methods to the full. Yet, given the global structural change towards service sector economies and the growing importance of tourism within them, further contributions towards explanation, prediction and quantification of tourism phenomena are necessary. The onus on economists is to demonstrate the relevance of their subject's concepts and methods to both theoretical and empirical analyses of tourism. Equally, those in other disciplines should take account of economic perspectives.

REFERENCES

Aaronovitch, S. and Sawyer, M. (1975) *Big Business*, London: Macmillan.

Abowd, J. and Card, D.A. (1989) 'On the covariance structure of earnings and hours changes', *Econometrica*, 57: 411–445.

Adams, P.D. and Parmenter, B.R. (1995) 'An applied general equilibrium analysis of the economic effects of tourism in a quite small, quite open economy', *Applied Economics*, 27, 10: 985–994.

Adkins, L. (1995) *Gendered Work: Sexuality, Family and the Labour Market*, Buckingham and Philadelphia: Open University Press.

Aghion, P. and Howitt, P. (1992) 'A model of growth through creative destruction', *Econometrica*, 60, 2: 323–351.

—— (1995) 'Research and development in the growth process', *Journal of Economic Growth*, 1, 1: 49–74.

Aislabie, C.J. (1988a) 'Tourism issues in developing countries', in C.A. Tisdell, C.J. Aislabie and P.J. Stanton (eds) *Economics of Tourism: Case Study and Analysis*, University of Newcastle, New South Wales: Institute of Industrial Economics, 346–378.

—— (1988b) 'Economics and tourism: major issues in the literature', in C.A. Tisdell, C.J. Aislabie and P.J. Stanton (eds) *Economics of Tourism: Case Study and Analysis*, University of Newcastle, New South Wales: Institute of Industrial Economics, 15–38.

Alam, A. (1995) 'The new trade theory and its relevance to the trade policies of developing countries', *The World Economy*, 23, 8: 367–385.

Anderson, L.G. (1995) 'Privatizing open access fisheries: individual transferable quotas', in D.W. Bromley (ed.) *Handbook of Environmental Economics*, Oxford: Blackwell, 453–474.

Andronikou, A. (1987) 'Cyprus: management of the tourist sector', *Tourism Management*, 7, 2: 127–129.

Archer, B. (1973) *The Impact of Domestic Tourism*, Occasional Papers in Economics, no. 2, Bangor: University of Wales Press.

Archer, B.H. (1976) *Demand Forecasting in Tourism*, Occasional Papers in Economics, no. 9, Bangor: University of Wales Press.

—— (1977a) *Tourism Multipliers: The State of the Art*, Occasional Papers in Economics, no. 11, Bangor: University of Wales Press.

—— (1977b) *Tourism in the Bahamas and Bermuda: Two Case Studies*, Occasional Papers in Economics, no. 10, Bangor: University of Wales Press.

—— (1989) 'Tourism and island economies: impact analyses', in C.P. Cooper (ed.) *Progress in Tourism, Recreation and Hospitality Management*, vol 1, London: Belhaven.

—— (1995) 'Importance of tourism for the economy of Bermuda', *Annals of Tourism Research*, 22, 4: 918–930.

Archer, B.H. and Fletcher, J. (1996) 'The economic impact of tourism in the Seychelles', *Annals of Tourism Research*, 23, 1: 32–47.

Arrow, K.J. (1962) 'The economic implications of learning by doing', *Review of Economic Studies*, 29: 155–173.

—— (1975) 'Vertical integration and communication', *Bell Journal of Economics*, 6: 173–183.

Artus, J.R. (1972) 'An econometric analysis of international travel', *IMF Staff Papers*, 19: 579–614.

Asabere, P.K., Hachey, G. and Grubaugh, S. (1989) 'Architecture, historic zoning, and the value of homes', *Journal of Real Estate, Finance and Economics*, 2: 181–195.

Asian Development Bank (ADB) (1995) *Indonesia–Malaysia–Thailand Growth Triangle Development Project*, Regional Technical Assistance 5550, vol. VI Tourism, Manila: Asian Development Bank.

Ayres, R.U. and Kneese, A.V. (1989) 'Externalities: economies and thermodynamics', in F. Archibugi and P. Nijkamp (eds) *Economy and Ecology: Towards Sustainable Development*, Dordrecht: Kluwer.

Bachmann, P. (1988) *Tourism in Kenya: Basic Need for Whom?*, Berne: Peter Lang.

Bagguley, P. (1990) 'Gender and labour flexibility in hotels and catering', *Service Industries Journal*, 10: 737–747.

Bain, J.S. (1956) *Barriers to New Competition*, Cambridge, Mass.: Harvard University Press.

Balchin, P.N., Kieve, J.L. and Bull, G.H. (1988) *Urban Land Economics and Public Policy*, 4th Edn, London: Macmillan.

Barbier, E.B. (1992) 'Community-based development in Africa', in T.M. Swanson and E.B. Barbier (eds) *Economics for the Wilds*, London: Earthscan.

Baretje, R. (1982) 'Tourism's external account and the balance of payments', *Annals of Tourism Research*, 9, 1: 57–67.

—— (1987) 'La contribution nette du tourisme international a la balance des paiements', *Problems of Tourism*, 10, 4: 51–88.

—— (1988) 'Tourisme international de tiers monde l'enjeu des devises', *Teoros*, 7, 3: 10–14.

Barnett, H. (1979) 'Scarcity and growth revisited', in V.K. Smith (ed.) *Scarcity and Growth Reconsidered*, Baltimore, Md: Johns Hopkins University Press.

Barnett, H. and Morse, C. (1963) *Scarcity and Growth: The Economics of Natural Resource Availability*, Baltimore: Johns Hopkins University Press.

BarOn, R.R. (1979) 'Forecasting tourism – theory and practice', TTRA (Travel and Tourism Research Association) Tenth Annual Conference Proceedings, University of Utah.

—— (1983) 'Forecasting tourism by means of travel series over various time spans under specified scenarios', Third International Symposium on Forecasting.

Barro, R.J. (1990) 'Government spending in a simple model of endogenous growth', *Journal of Political Economy*, 98: S103–S125.

—— (1991) 'Economic growth in a cross-section of countries', *Quarterly Journal of Economics*, 106: 409–443.

Barro, R.J. and Sala-i-Martin, X. (1992) 'Public finance in models of economic growth', *Review of Economic Studies*, 54: 646–661.

Barry, K. and O'Hagan, K. (1971) 'An econometric study of British tourist expenditure in Ireland', *Economic and Social Review*, 3, 2: 143–161.

Basu, K. (1993) *Lectures in Industrial Organization Theory*, Oxford: Basil Blackwell.

Bateman, I.J., Willis, K.G., Garrod, G.D., Doktor, P., Langford, I. and Turner, R.K. (1992) 'Recreation and environmental preservation value of the Norfolk Broads:

a contingent valuation study', Unpublished Report, Environmental Appraisal Group, University of East Anglia.

Bateman, I.J., Willis, K.G. and Garrod, G. (1994) 'Consistency between contingent valuation estimates: a comparison of two studies of UK National Parks', *Regional Studies*, 28, 5: 457–474.

Baum, T. and Mudambi, R. (1995) 'An empirical analysis of oligopolistic hotel pricing', *Annals of Tourism Research*, 22, 3: 501–516.

Baumol, W.J. (1982) 'Contestable markets: an uprising in the theory of industry structure', *American Economic Review*, 72: 1–15.

Bechdolt Jr, B.V. (1973) 'Cross-sectional travel demand functions: US visitors to Hawaii, 1961–70', *Quarterly Review of Economics and Business*, 13: 37–47.

Beioley, S. (1995) 'Green tourism – soft or sustainable?', *Insights*, May: B75–B89.

Bennett, M.M. (1993) 'Information technology and travel agency: a customer service perspective', *Tourism Management*, 14,4: 259–66.

Berkes, F. (ed.) (1989) *Common Property Resources: Ecology and Community-Based Sustainable Development*, London: Belhaven.

Bird, R.M. (1992) 'Taxing tourism in developing countries', *World Development*, 20, 1145–1158.

Blackwell, D. (1996) 'Airtours chief will cut stake to aid carnival', *Financial Times*, 20 March.

Blinder, A.S. and Deaton, A.S. (1985) 'The time-series consumption revisited', *Brookings Papers on Economic Activity*, 465–521.

Blundell, R. (1991) 'Consumer behaviour: theory and empirical evidence – a survey', in A.J. Oswald (ed.) *Surveys in Economics*, vol. 2, Oxford: Basil Blackwell.

Blundell, R., Pashardes, P. and Weber, G. (1993) 'What do we learn about consumer demand patterns from micro data?', *American Economic Review*, 83, 3: 570–597.

Board, J., Sinclair, M.T. and Sutcliffe, C.M.S. (1987) 'A portfolio approach to regional tourism', *Built Environment*, 13, 2: 124–137.

Bockstael, N.E., McConnell, K.E. and Strand, I.E. (1991) 'Recreation', in J.B. Braden and C.D. Kolstad (eds) *Measuring the Demand for Environmental Quality*, New York: North-Holland.

Bote Gómez, V. (1988) *Turismo en espacio rural: rehabilitación del patrimonio sociocultural y de la economía local*, Madrid: Editorial Popular.

—— (1990) *Planificación económica del turismo: de una estrategia masiva a una artesanal*, Mexico: Editorial Trillas.

—— (1993) 'La necesaria revalorización de la actividad turística española en una economía terciarizada e integrada en la CEE', *Estudios Turísticos*, no. 118: 5–26.

Bote Gómez, V. and Sinclair, M.T. (1991) 'Integration in the tourism industry', in M.T. Sinclair and M.J. Stabler (eds) *The Tourism Industry: An International Analysis*, Wallingford: CAB International.

—— (1996) 'Tourism demand and supply in Spain', in M. Barke, M. Newton and J. Towner (eds) *Tourism in Spain: Critical Perspectives*, Wallingford: C.A.B. International.

Bote Gómez, V., Sinclair, M.T., Sutcliffe, C.M.S. and Valenzuela, M. (1989) 'Vertical integration in the British/Spanish tourism industry', in *Leisure, Labour and Lifestyles: International Comparisons, Tourism and Leisure. Models and Theories*, Proceedings of the Leisure Studies Association Second International Conference, Brighton, Conference Papers no. 39, 8, 1: 80–96.

Bote Gómez, V., Huescar, A. and Vogeler, C. (1991) 'Concentracion e integración de las agencias de viajes españolas ante el Acta Unica Europea', *Papers de Turisme*, no. 5, 5–43, Instituto Turístico Valenciano, Valencia.

Boulding, K.E. (1966) 'The Economics of the coming Spaceship Earth', in H. Jarrett (ed.) *Environmental Quality in a Growing Economy*, Baltimore: Johns Hopkins University Press.

Brander, J.A. (1981) 'Intra-industry trade in identical commodities', *Journal of International Economics*, 11: 1–14.

Brander, J.A. and Krugman, P.R. (1983) 'A reciprocal dumping model of international trade', *Journal of International Economics*, 15: 313–321.

Braun, P.A., Constantinides, G.M. and Ferson, W.E. (1993) 'Time nonseparability in aggregate consumption: international evidence', *European Economic Review*, 37, 5: 897–920.

Breathnach, P., Henry, M., Drea, S. and O'Flaherty, M. (1994) 'Gender in Irish tourism employment', in V. Kinnaird and D. Hall (eds) *Gender: A Tourism Analysis*, Chichester: John Wiley.

Brierton, U.A. (1991) 'Tourism and the environment', *Contours*, 5, 4: 18–19.

Britton, S.G. (1980) 'A conceptual model of tourism in a peripheral economy', in D.G. Pearce (ed.) *Tourism in the South Pacific: the Contribution of Research to Development and Planning*, Christchurch: University of Canterbury.

—— (1982) 'The political economy of tourism in the third world', *Annals of Tourism Research*, 9, 3: 331–358.

Brockhoff, K. (1975) 'The performance of forecasting groups in computer dialogue and face to face discussion', in H.A. Limestone, and M. Turoff (eds) *The Delphi Method Techniques and Applications*, Reading, Mass.: Addison-Wesley.

Brooks, C., Cheshire, P.C., Evans, A.W. and Stabler, M.J. (1995) *The Economic and Social Value of the Conservation of Historic Buildings and Areas*, Report prepared for English Heritage, Department of National Heritage and Royal Institution of Chartered Surveyors, University of Reading: Department of Economics.

Brookshire, D., Eubanks, L. and Randall A. (1983) 'Estimating option price and existence values for wildlife resources', *Land Economics,* 59, 1: 1–15.

Brown, G. Jr., and Henry, W. (1989) *The Economic Value of Elephants*, London Environmental Economics Centre Paper 89–12, University College London.

Brozen, Y. (1971) 'Bain's concentration and rates of return revisited', *Journal of Law and Economics*, 14: 351–369.

Bryden, J.M. (1973) *Tourism and Development: A Case Study of the Commonwealth Caribbean*, Cambridge: Cambridge University Press.

Buchanan, J.M. (1968) *Demand and Supply of Public Goods*, Chicago, Ill.: Rand McNally.

Buckley, P.J. (1987) 'Tourism: an economic transactions analysis', *Tourism Management*, 8: 190–204.

Bull, A. (1991) *The Economics of Travel and Tourism*, Melbourne: Pitman.

Burkart, A.J. and Medlik, S. (1989) *Tourism: Past, Present and Future*, 2nd edn, London: Heinemann.

Burns, A.C. and Ortinau, D.J. (1979) 'Underlying perceptual patterns in husband and wife purchase decision influence assessments', *Advances in Consumer Research*, 6: 372–376.

Burns, P. and Cleverdon, R. (1995) 'Destination on the edge? The case of the Cook Islands', in M.V. Conlin and T. Baum (eds) *Island Tourism*, Chichester: John Wiley.

Burns, P. and Holden, A. (1995) *Tourism: A New Perspective*, London: Prentice Hall.

Burt, O. and Brewer, D. (1974) 'Estimation of net social benefits from outdoor recreation', *Econometrica*, 39: 813–827.

Butler, R.W. (1980) 'The concept of a tourist area cycle of evolution: implications for management of resources', *Canadian Geographer*, 14: 5–12.

Buttle, F. (1988) *Hotel and Food Science Marketing: A Managerial Approach*, London: Cassell.

Button, K. (ed.) (1991) *Airline Deregulation*, London: David Fulton.

Caballero, R.J. (1993) 'Durable goods: an explanation for their slow adjustment', *Journal of Political Economy*, 101, 2: 351–384.

Campbell, C.K. (1967) 'An approach to research in recreational geography', British Columbia Occasional Papers no. 7, Department of Geography, Vancouver: University of British Columbia.

Campbell, J.Y. and Mankiw, N.G. (1991) 'The response of consumption to income: a cross-country investigation', *European Economic Review*, 35: 715–721.

Carlton, D.W. (1979) 'Vertical integration in competitive markets under uncertainty', *Journal of Industrial Economics*, 27: 189–209.

Carson, R. (1963) *The Silent Spring*, London: Hamish Hamilton.

Casson, M.C. (1987) *The Firm and the Market*, Oxford: Basil Blackwell.

Castelberg-Koulma, M. (1991) 'Greek women and tourism: women's co-operatives as an alternative form of organization', in N. Redclift and M.T. Sinclair (eds) *Working Women: International Perspectives on Labour and Gender Ideology*, London and New York: Routledge.

Cater, E. (1993) 'Ecotourism in the third world: problems for sustainable tourism development', *Tourism Management*, 14, 2: 85–90.

Cater, E. and Goodall, B. (1992) 'Must tourism destroy its resource base?', in A.M. Mannion and S.R. Bowlby (eds) *Environmental Issues in the 1990s*, Chichester: John Wiley.

Cater, E. and Lowman, G. (eds) (1994) *Ecotourism: A Sustainable Option?*, Chichester: John Wiley.

Cazes, G. (1972) 'Le rôle du tourisme dans la croissance économique: reflexions a partir de trois examples antillais', *The Tourist Review*, 27, 3: 93–98 and 144–148.

Chadha, B. (1991) 'Wages, profitability and growth in a small open economy', *IMF Staff Papers*, 38, 1: 59–82.

Chakwin, N. and Hamid, N. (1996) 'The economic environment in Asia for investment', in C. Fry and C. Oman (eds) *Investing in Asia*, Paris: Organization for Economic Cooperation and Development.

Chamberlin, E.H. (1933) *The Theory of Monopolistic Competition*, Cambridge, Mass.: Harvard University Press.

Chant, S. (1997) 'Gender and tourism employment in Mexico and the Philippines', in M.T. Sinclair (ed.) *Gender, Work and Tourism*, London and New York: Routledge.

Cheshire, P.C. and Sheppard, S. (1995) 'On the price of land and the value of amenities', *Economica*, 62: 247–268.

Cheshire, P.C. and Stabler, M.J. (1976) 'Joint consumption benefits in recreational site "surplus": an empirical estimate', *Regional Studies*, 10: 343–351.

Chevas, J.P., Stoll, J. and Sellar, C. (1989) 'On the commodity value of travel time in recreational activities', *Applied Economics*, 21: 711–722.

Civil Aviation Authority (CAA) (1994) *Business Monitor*, July.

Clark, C.W. (1973) 'The economics of overexploitation', *Science*, 181: 630–634.

—— (1990) *Mathematical Bioeconomics: The Optimal Management of Renewable Resources*, 2nd edn, New York: Wiley.

Clarke, C.D. (1981) 'An analysis of the determinants of the demand for tourism in Barbados', Ph.D. thesis, Fordham University, USA.

Clarke, R. and Davies, S.W. (1982) 'Market structure and price-cost margins', *Economica*, 49: 277–288.

Clawson, M. and Knetsch, J.L. (1966) *Economics of Outdoor Recreation*, Baltimore: Johns Hopkins University Press.

Cleverdon, R. and Edwards, E. (1982) *International Tourism to 1990*, Cambridge: Abt Books.

Clewer, A., Pack, A. and Sinclair, M.T. (1990) 'Forecasting models for tourism demand in city-dominated and coastal areas', *European Papers of the Regional Science Association*, 69: 31–42.

—— (1992) 'Price competitiveness and inclusive tour holidays in European cities', in P. Johnson and B. Thomas (eds) *Choice and Demand in Tourism*, London: Mansell.

227

REFERENCES

Coase, R.H. (1937) 'The nature of the firm', *Economica* (new series), 4: 386–405.

Coase, R. (1960) 'The problem of social cost', *Journal of Law and Economics*, 3: 1–44.

Cohen, E. (1978) 'Impact of tourism on the physical environment', *Annals of Tourism Research*, 5, 2: 215–237.

Combs, J.P., Kirkpatrick, R.C., Shogren, J.F. and Herriges, J.A.(1993) 'Matching grants and public goods: a closed-ended contingent valuation experiment', *Public Finance Quarterly*, 21, 2: 178–195.

Commission of the European Communities (1994) *Report by the Commission to the Council, to the European Parliament and the Economic and Social Committee on Community Measures affecting Tourism*, Council Decision 92/421/EEC, Brussels: Commission of the European Communities.

—— (1995a) *The Role of the Union in the Field of Tourism*, Commission Green Paper, COM (95)97, Brussels: Commission of the European Communities.

—— (1995b) *Consultation on the basis of the Green Paper: A Step further towards Recognition of Community Action to Assist Tourism, Forum on European Tourism*, Brussels: Commission of the European Communities.

—— (1996) *Proposal for a Council Decision on a First Multiannual Programme to Assist European Tourism 'Philoxenia' (1997–2000)*, COM(96)168, Brussels: Commission of the European Communities.

Common, M. (1973) 'A note on the use of the Clawson method', *Regional Studies*, 7: 401–406.

Conlin, M.V. and Baum, T. (1995) *Island Tourism*, Chichester: John Wiley.

Conrad, J.M. (1995) 'Bioeconomic models of the fishery', in D.W. Bromley (ed.) *Handbook of Environmental Economics*, Oxford: Blackwell.

Cook, S.D., Stewart, E. and Repass, K. (1992) *Discover America: Tourism and the Environment*, Washington, DC: Travel Industry Association of America.

Cooper, C.P. (ed.) (1989) *Progress in Tourism, Recreation and Hospitality Management*, vol. 1, London: Belhaven.

—— (ed.) (1990) *Progress in Tourism, Recreation and Hospitality Management*, vol. 2, London: Belhaven.

—— (ed.) (1991) *Progress in Tourism, Recreation and Hospitality Management*, vol. 3, London: Belhaven.

—— (1992) 'The life cycle concept and tourism', in P. Johnson and B. Thomas (eds) *Choice and Demand in Tourism*, London: Mansell.

Cooper, C.P. and Lockwood, A. (eds) (1992) *Progress in Tourism, Recreation and Hospitality Management*, vol. 4, London: Belhaven.

Cooper, C.P., Fletcher, J., Gilbert, D. and Wanhill, S. (1993) *Tourism: Principles and Practice*, London: Pitman.

Copeland, B.R. (1991) 'Tourism, welfare and de-industrialization in a small open economy', *Economica*, 58, 4: 515–529.

Cowling, K. and Waterson, M. (1976) 'Price-cost margins and market structure', *Economica*, 43: 267–274.

Crompton, J.L. (1979) 'Motivations for pleasure vacation', *Annals of Tourism Research*, 6, 4: 408–424.

Crutchfield, J.A. and Zellner, A. (1962) 'Economic aspects of the Pacific halibut fishery', *Fisher Industrial Research*, 1, 1.

Curry, S. (1982) 'The terms of trade and real import capacity of the tourism sector in Tanzania', *Journal of Development Studies*, 18, 4: 479–496.

Curry, S. and Morvaridi, B. (1992) 'Sustainable tourism: illustrations from Kenya, Nepal and Jamaica', in C.P. Cooper and A. Lockwood (eds) *Progress in Tourism, Recreation and Hospitality Management*, vol. 4, London: Belhaven.

Cyert, R.M. and March, J.G. (1963) *A Behavioural Theory of the Firm*, Englewood Cliffs, NJ: Prentice Hall.

Cyert, R.M. and Simon, H.A. (1983) 'The behavioural approach: with emphasis on economics', *Behavioural Science*, 28: 95–108.

Dalkey, N. and Helmer, O. (1963) 'An experimental application of the Delphi method of the use of experts', *Management Science*, 9, 3: 458–467.

Daly, H. (1977) *Steady State Economics*, San Francisco: Freeman.

D'Amore, L.J. (1992) 'Promoting sustainable tourism: the Canadian approach', *Tourism Management*, 13, 3: 258–262.

Daneshkhu, S. (1996) 'Inter-Continental gains four hotels in Malaysia', *Financial Times*, 9 September.

—— (1997) 'Travel agents join forces', *Financial Times*, 13 January.

Dann, G.M.S. (1981) 'Tourist motivation: an appraisal', *Annals of Tourism Research*, 8, 2: 187–219.

Dardin, R. (1980) 'The value of life: new evidence from the marketplace', *American Economic Review*, 70: 1077–1082.

Darnell, A., Johnson, P. and Thomas, B. (1992) 'Modelling visitor flows at the Beamish Museum', in P. Johnson and B. Thomas (eds) *Choice and Demand in Tourism*, London: Mansell.

Dasgupta, P. and Stiglitz, J. (1980) 'Industrial structure and the nature of innovative activity', *Economic Journal*, 90: 266–293.

Davidson, J.E., Hendry, D.F., Srba, F. and Yeo, S. (1978) 'Econometric modelling of the aggregate time-series relationship between consumers' expenditure and income in the United Kingdom', *Economic Journal*, 88: 661–92.

Davies, S. (1989a) 'Concentration', in S. Davies *et al.* (eds) *Economics of Industrial Organisation*, London and New York: Longman.

—— (1989b) 'Technical change, productivity and market structure', in S. Davies *et al.* (eds) *Economics of Industrial Organisation*, London and New York: Longman.

Davies, S., Lyons, B. with Dixon, H. and Gerowski, P. (1989) *Economics of Industrial Organisation*, London and New York: Longman.

Davis, H.L. (1970) 'Dimensions of marital roles in consumer decision-making', *Journal of Marketing Research*, 7: 168–177.

Davis, O.A. and Whinston, A.B. (1961) 'The economics of urban renewal', in J.Q. Wilson (ed.) *Urban Renewal: The Record and the Controversy*, Cambridge, Mass.: MIT Press.

Deaton, A.S. (1992) *Understanding Consumption*, Oxford: Clarendon Press.

Deaton, A.S. and Muellbauer, J. (1980a) 'An almost ideal demand system', *American Economic Review*, 70, 3: 312–26.

—— (1980b) *Economics and Consumer Behaviour*, Cambridge: Cambridge University Press.

Debbage, K.G. (1990) 'Oligopoly and the resort cycle in the Bahamas', *Annals of Tourism Research*, 17, 4: 513–527.

de Kadt, E. (1979) *Tourism: Passport to Development*, Oxford: Oxford University Press.

Delos Santos, J.S., Ortiz, E.M., Huang, E. and Secretario, F. (1983) 'Philippines', in E.A. Pye and T-b. Lin (eds) *Tourism in Asia: The Economic Impact*, Singapore: Singapore University Press.

de Mello Jr, L.R. and Sinclair, M.T. (1995) *Foreign Direct Investment, Joint Ventures and Endogenous Growth*, Studies in Economics no. 95/13, University of Kent at Canterbury.

Demsetz, H. (1969) 'Information and efficiency: another viewpoint', *Journal of Law and Economics*, 12, 1: 1–22.

—— (1974) 'Two systems of belief about monopoly', in H.J. Goldschmid, H.M. Mann and J.F. Weston (eds), *Industrial Concentration: The New Learning*, Boston, Mass.: Little, Brown.

Desvousges, W.S., Smith, V.K. and McGivney, M.P. (1983) *Comparison of Alternative Approaches for Estimating Recreation and Related Benefits of Water Quality Improvements*, US Environmental Protection Agency, EPA-230–05–83–001, Washington, DC.

Dharmaratne, G.S. (1995) 'Forecasting tourist arrivals in Barbados', *Annals of Tourism Research*, 22, 4: 804–818.

Diamond, D. and Richardson, R. (1996) *The Economic Significance of the British Countryside*, London: The Countryside Business Group.

Diamond, J. (1974) 'International tourism and the developing countries: a case study in failure', *Economica Internazionale*, 27, 3–4: 601–615.

—— (1977) 'Tourism's role in economic development: the case reexamined', *Economic Development and Cultural Change*, 25, 3: 539–53.

di Benedetto, C.A. and Bojanic, D.C. (1993) 'Tourism area life cycle extensions', *Annals of Tourism Research*, 20, 3: 557–570.

Dickie, M. and Gerking, S. (1991) 'Willingness to pay for ozone control: inferences from the demand for medical care', *Journal of Environmental Economics and Management*, 21; 1–16.

Dieke, P.U.C. (1993a) 'Tourism and development policy in The Gambia', *Annals of Tourism Research*, 20, 3: 423–449.

—— (1993b) 'Tourism in The Gambia: some issues in development policy', *World Development*, 21, 2: 277–289.

—— (1995) 'Tourism and structural adjustment programmes in the African economy', *Tourism Economics*, 1: 71–93.

Dietrich, M. (1994) *'Transaction Cost Economics and Beyond: Towards a New Economics of the Firm'*, Routledge: London.

Dingle, P.A.J.M. (1995) 'Practical green business', *Insights*, March: C35–C45.

Dixit, A.K. (1982) 'Recent developments in Oligopoly Theory', *American Economic Review Papers and Proceedings*, 72: 12–17.

Domar, E.D. (1946) 'Capital expansion, rate of growth and employment', *Econometrica*, 14: 137–147.

—— (1947) 'Expansion and employment', *American Economic Review*, 37, 1: 34–55.

Douglas, N. (1997) 'Applying the life cycle model to Melanesia', *Annals of Tourism Research*, 24, 1: 1–22.

Drexl, C. and Agel, P. (1987) 'Tour operators in West Germany: survey of the package tour market, the operators and how they sell', *Travel and Tourism Analyst*, May: 29–43.

Drobny, A. and Hall, S.G. (1989) 'An investigation of the long run properties of aggregate non-durable consumers' expenditure in the United Kingdom', *Economic Journal*, 99: 454–460.

Duesenberry, J.S. (1949) *Income, Saving and the Theory of Consumer Behavior*, Cambridge, Mass.: Harvard University Press.

Dunning, J.H. and McQueen, M. (1982a) *Transnational Corporations in International Tourism*, New York: United Nations Centre for Transnational Corporations.

—— (1982b) 'Multinational corporations in the international hotel industry', *Annals of Tourism Research*, 9, 1: 69–90.

Dwyer, L. and Forsythe, P. (1994) 'Foreign tourism investment: motivation and impact', *Annals of Tourism Research*, 21, 3: 512–537.

Eadington, W.R. and Redman, M. (1991) 'Economics and tourism', *Annals of Tourism Research*, 18, 1: 41–56.

East, M. (1994) 'Second tier operators strengthen the industry', *Travel Weekly*, 12 January.

Eber, S. (ed.) (1992) *Beyond the Green Horizon: Principles for Sustainable Tourism*, Godalming: World Wide Fund for Nature.

Economist, The (1996) 'Go to dreamland, forget the mosques', *The Economist,* 17 August.

Eggertson, T. (1990) *Economic Behaviour and Institutions,* New York: Cambridge University Press.

Elkington, J. and Hailes, J. (1992) *Holidays That Don't Cost the Earth,* London: Victor Gollancz.

Englin, J. and Mendelsohn, R. (1991) 'A hedonic travel cost analysis for valuation of multiple components of site quality: the recreational value of forest management', *Journal of Environmental Economics and Management,* 21: 275–290.

English, E.P. (1986) *The Great Escape? An Examination of North–South Tourism,* Ottawa: North–South Institute.

Evans, A.W.E. (1982) 'Externalities, rent seeking and town planning', Discussion Papers in Urban and Regional Economics no. 10, University of Reading: Department of Economics.

—— (1985) *Urban Economics. An Introduction,* Oxford: Basil Blackwell.

Evans, N. and Stabler, M.J. (1995) 'A future for the package tour operator in the 21st century?', *Tourism Economics,* 1, 3: 245–263.

Fairbairn-Dunlop, P. (1994) 'Gender, culture and tourism development in Western Samoa', in V. Kinnaird and D. Hall (eds) *Tourism: A Gender Analysis,* Chichester: John Wiley.

Farver, J.A.M. (1984) 'Tourism and employment in The Gambia', *Annals of Tourism Research,* 11, 2: 249–265.

Ferguson, P.R. and Ferguson, G.J. (1994) *Industrial Economics: Issues and Perspectives,* 2nd edn, London: Macmillan.

Figuerola Palomo, M. (1991) *Elementos para el Estudio de la Economía de la Empresa Turística,* Madrid: Editorial Sistesis.

Filiatrault, P. and Ritchie, J.R.B. (1980) 'Joint purchasing decisions: a comparison of influence structure in family and couple decision-making unit', *Journal of Consumer Research,* 7: 131–140.

Fish, M. (1982) 'Taxing international tourism in West Africa', *Annals of Tourism Research,* 9, 1: 91–103.

Fishbein, M. (1963) 'An investigation of the relationships between beliefs about an object and the attitude toward that object', *Human Relationships,* 16: 233–240.

Fitch, A. (1987) 'Tour operators in the UK: survey of the industry, its markets and product diversification', *Travel and Tourism Analyst,* March: 29–43.

Flavin, M. (1981) 'The adjustment of consumption to changing expectations about future income', *Journal of Political Economy,* 89: 974–1009.

Fletcher, J.E. (1986a) *The Economic Impact of International Tourism on the National Economy of the Republic of Palau,* Madrid: World Tourism Organization/United Nations Development Programme.

—— (1986b) *The Economic Impact of Tourism on Western Samoa,* Madrid: World Tourism Organization/United Nations Development Programme.

—— (1987) *The Economic Impact of International Tourism on the National Economy of the Solomon Islands,* Madrid: World Tourism Organization/United Nations Development Programme.

Fletcher, J.E. and Archer, B.H. (1991) 'The development and application of multiplier analysis', in C.P. Cooper (ed.) *Progress in Tourism, Recreation and Hospitality Management,* vol. 1, London: Belhaven.

Ford, D.A. (1989) 'The effect of historic district designation on single-family home prices', *AREUEA Journal,* 17, 3: 353–362.

Forrester, J.W. (1971) *World Dynamics,* Cambridge, Mass.: Allen Press.

Forsyth, T. (1995) 'Business attitudes to sustainable tourism: responsibility and self regulation in the UK outgoing tourism industry', Paper presented at the Sustainable Tourism World Conference, Lanzarote.

Foster, D. (1985) *Travel and Tourism Management*, London: Macmillan.

Freeman, C. (1979) 'Technical innovation and British trade performance', in F. Blackaby (ed.) *Deindustrialisation*, London: Heinemann International.

Friedman, M. (1957) *A Theory of the Consumption Function*, Princeton: Princeton, NJ: University Press.

—— (1968) 'The role of monetary policy', *American Economic Review*, 38: 1–17.

Fritz, R.G., Brandon, C. and Xander, J. (1984) 'Combining the time-series and econometric forecast of tourism activity', *Annals of Tourism research*, 11, 2: 219–229.

Fujii, E., Khaled, M. and Mak, J. (1985) 'The exportability of hotel occupancy and other tourist taxes', *National Tax Journal*, 38, 2: 169–177.

—— (1987) 'An empirical comparison of systems of demand equations: an application to visitor expenditure in resort destinations', *Philippine Review of Business and Economics*, 24, 1–2: 79–102.

Garrod, G. and Willis, K. (1990) 'Contingent valuation techniques: a review of their unbiasedness, efficiency and consistency', *Countryside Change Unit Working Paper*, no. 10, Newcastle-upon-Tyne: University of Newcastle.

—— (1991a) 'The environmental economic impact of woodland: a two-stage hedonic price model of the amenity value of forestry in Britain', *Applied Economics*, 24: 715–728.

—— (1991b) 'Some empirical estimates of forest amenity value', *Countryside Change Unit Working Paper*, no. 13. Newcastle-upon-Tyne: University of Newcastle.

—— (1991c) 'The hedonic price method and the valuation of the countryside characteristics', *Countryside Change Unit Working Paper*, no. 14. Newcastle-upon-Tyne: University of Newcastle.

Garrod, G., Pickering, A. and Willis, K. (1991) 'An economic estimation of the recreational benefit of four botanic gardens', *Countryside Change Unit Working Paper*, no. 25, Newcastle-upon-Tyne: University of Newcastle.

Garrod, G., Willis, K. and Saunders, C.M. (1994) 'The benefits and costs of the Somerset levels and moors ESA', *Journal of Rural Studies*, 10, 2: 131–145.

Getz, D. (1986) 'Models in tourism planning', *Tourism Management*, 7, 1: 21–32.

—— (1992) 'Tourism planning and destination life cycle', *Annals of Tourism Research*, 19, 4: 752–770.

Geurts, M.D. (1982) 'Forecasting the Hawaiian tourist market', *Journal of Travel Research*, 11, 1: 18–21.

Geurts, M.D. and Ibrahim, I.B. (1975) 'Comparing the Box-Jenkins approach with the exponentially smoothed forecasting model: application to Hawaii tourists', *Journal of Marketing Research*, 12: 182–188.

Global Environmental Management Initiative (GEMI) (1992) '*Environmental self-assessment based on the International Chamber of Commerce's Business Charter for Sustainable Development*', Washington, DC: GEMI.

Go, F. (1988) 'Key problems and prospects in the international hotel industry', *Travel and Tourism Analyst*, 1: 27–49.

—— (1989) 'International hotel industry: capitalizing on change', *Tourism Management*, 10, 3: 195–200.

Godbey, G. (1988) 'Play as a model for the study of tourism', Paper presented at the Leisure Studies Association 2nd International Conference, Leisure, Labour and Lifestyles: International Comparisons, June, Brighton: University of Sussex.

Gonzalez, P. and Moral, P. (1996) 'Analysis of tourism trends in Spain', *Annals of Tourism Research*, 23, 4: 739–754.

REFERENCES

Goodall, B. (1987) 'Tourism and jobs in the United Kingdom', *Built Environment*, 15, 2: 78–91.

—— (1992) 'Environmental auditing for tourism', in C.P. Cooper and A. Lockwood (eds) *Progress in Tourism, Recreation and Hospitality Management*, vol. 4, London: Belhaven.

Goodall, B. and Ashworth, G.J. (eds) (1988) *Marketing in the Tourism Industry*, Beckenham: Croom Helm.

Goodall, B. and Stabler, M.J. (1994) 'Tourism-environment issues and approaches to their solution', in H. Voogd (ed.) *Issues in Environmental Planning*, London: Pion.

Gordon, H.S. (1954) 'The economic theory of a common property resource: the fishery', *Journal of Political Economy*, 62: 124–142.

Gordon, I.R. (1994) 'Crowding, competition and externalities in tourism development: a model of resort life cycles', *Geographical Systems*, 1: 289–308.

Gordon, I.R. and Goodall, B. (1992) 'Resort cycles and development processes', *Built Environment*, 18: 41–56.

Graburn, N.H.H. (1983) 'Tourism and prostitution', *Annals of Tourism Research*, 10: 437–443.

Gray, H.P. (1966) 'The demand for international travel by the United States and Canada', *International Economic Review*, 7, 1: 83–92.

—— (1970) *International Travel – International Trade*, Lexington, Mass.: D.C. Heath.

—— (1982) 'The contributions of economics to tourism', *Annals of Tourism Research*, 9, 1: 105–125.

—— (1984) 'Tourism theory and practice: a reply to Alberto Sessa', *Annals of Tourism Research*, 11: 286–289.

Green, H. and Hunter, C. (1992) 'The environmental impact assessment of tourism development', in P. Johnson and B. Thomas (eds) *Perspectives on Tourism Policy*, London: Mansell.

Green, H., Hunter, C. and Moore, B. (1990a) 'Application of the Delphi technique in tourism', *Annals of Tourism Research*, 17, 2: 270–279.

—— (1990b), 'Assessing the environmental impact of tourism development: Use of the Delphi technique', *Tourism Management*, 11, 2: 111–120.

Greenaway, D., Bleaney, M. and Stewart, I. (1996) *Guide to Modern Economics*, London and New York: Routledge.

Greenwood, D.J. (1976) 'Tourism as an agent of change', *Annals of Tourism Research*, 3: 128–142.

Grinstein, A. (1955) 'Vacations: a psycho-analytic study', *International Journal of Psycho-Analysis*, 36, 3: 177–185.

Grossman, G.M. and Helpman, E. (1990a) 'Comparative advantage and long run growth', *American Economic Review*, 80: 796–815.

—— (1990b) 'Trade, innovation, and growth', *American Economic Review*, 80: 86–91.

—— (1991) *Innovation and Growth in the Global Economy*, Cambridge, Mass.: MIT Press.

—— (1994) 'Endogenous innovations in the theory of growth', *Journal of Economic Perspectives*, 8: 23–44.

Grünthal, A. (1960) 'Foreign travel in the balance of payments', *The Tourist Review*, 1: 14–20.

Guerrier, Y. (1986) 'Hotel manager: an unsuitable job for a woman?', *The Service Industries Journal*, 6, 2: 227–240.

Gunadhi, H. and Boey, C.K. (1986) 'Demand elasticities of tourism in Singapore', *Tourism Management*, 7, 4: 239–253.

Gunn, C.A. (1988) *Tourism Planning*, New York: Taylor & Francis.

Hall, D.C. and Hall, J.V. (1984) 'Concepts and measures of natural resource scarcity', *Journal of Environmental Economics and Management*, 11: 369–370.

Hanley, N. (1988) 'Using contingent valuation to value environmental improvements', *Applied Economics*, 20: 541–549.

—— (1989) 'Problems in valuing environmental improvements resulting from agricultural policy changes: the case of nitrate pollution', *Discussion Paper* no. 89/1, University of Stirling: Economic Department.

Hanley, N. and Ruffell, R. (1992) 'The valuation of forest characteristics', *Queen's Discussion Paper*, 849.

—— (1993) 'The contingent valuation of forest characteristics: two experiments', *Journal of Agricultural Economics*, 44: 218–229.

Hanley, N. and Spash, C.L. (1993) *Cost Benefit Analysis and the Environment*, Aldershot, Hants: Edward Elgar.

Hannah, L. and Kay, J.A. (1977) *Concentration in Modern Industry: Theory, Measurement and the UK Experience*, London: Macmillan.

Hansen, L.T. and Hallam, A. (1991) 'National estimates of the recreational value of streamflow', *Water Resources Research*, 27, 2: 167–175.

Harrigan, K.R. (1985) 'Exit barriers and vertical integration', *Academy of Management Journal*, 28: 686–697.

Harrison, A.J.M. and Stabler, M.J. (1981) 'An analysis of journeys for canal-based recreation', *Regional Studies* 15, 5 : 345–358.

Harrison, D. (1992) 'International tourism and the less developed countries: the background', in D. Harrison (ed.) *Tourism and the Less Developed Countries*, London: Belhaven.

—— (1995) 'Development of tourism in Swaziland', *Annals of Tourism Research*, 22, 1: 135–156.

Harrod, R.F. (1939) 'An essay in dynamic theory', *Economic Journal*, 49: 14–33.

Hawkins, D.E., Shafer, E.L. and Rovelstad, J.M. (eds) (1980) *Tourism Marketing and Management Issues*, Washington, DC: George Washington University.

Hay, D.A. and Morris, D.J. (1991) *Industrial Economics and Organization: Theory and Evidence*, Oxford: Oxford University Press.

Hayashi, F. (1985) 'The effect of liquidity constraints on consumption: a cross-sectional analysis', *Quarterly Journal of Economics*, 100: 183–206.

Hayek, F.A. (1949) *Individualism and Economic Order*, London: Routledge & Kegan Paul.

Haberlein, T. and Bishop, R. (1986) 'Assessing the validity of contingent valuation: three experiments', *Science of the Total Environment*, 56: 99–107.

Heckman, J.J. (1974) 'Life-cycle consumption and labor supply: an exploration of the relationship between income and consumption over the life cycle', *American Economic Review*, 64: 188–194.

Helpman, E. and Krugman, P.R. (1993) *Market Structures and Foreign Trade*, Cambridge, Mass.: MIT Press.

Helu Thaman, K. (1992) 'Beyond Hula, hotels and handicrafts', *In Focus*, 4, Summer: 8–9.

Hemming, C. (1993) *Business Success from Seizing the Environmental Initiative*, London: Business and the Environment Practitioner Series, Technical Communications (Publishing).

Henderson, D.M. and Cousins, R.L. (1975) *The Economic Impact of Tourism: A Case Study of Greater Tayside*, Tourism and Recreation Research Unit, Research Report no. 13, University of Edinburgh.

Hendry, D.F. (1983) 'Econometric modelling: the "consumption function" in retrospect', *Scottish Journal of Political Economy*, 30: 193–220.

Hendry, D.F. and Mizon, G.E. (1978) 'Serial correlation as a convenient simplification, not a nuisance', *Economic Journal*, 88: 549–563.

Heng, T.M. and Low, L. (1990) 'The economic impact of tourism in Singapore', *Annals of Tourism Research*, 17, 2: 246–269.

Hennessy, S. (1994) 'Female employment in tourism development in south-west England', in V. Kinnaird and D. Hall (eds) *Tourism: A Gender Analysis*, Chichester: John Wiley.

Hicks, L. (1990) 'Excluded women: how can this happen in the hotel world?', *The Service Industries Journal*, 10, 2: 348–363.

Hill, J. (1992) *Towards Good Environmental Practice: A Book of Case Studies*, London: Institute of Business Ethics.

Hill, J., Marshall, I. and Priddey, C. (1994) *Benefiting Business and the Environment: Case Studies of Cost Savings and New Opportunities from Environmental Initiatives*, London: Institute of Business Ethics.

Hirschman, A.O. (1964) 'The paternity of an index', *American Economic Review*, 54: 761–762.

Hirshleifer, J. (1982) *Research in Law and Economics*, vol. 4, *Evolutionary Models in Economics and Law*, London: JAI Press.

Hodgson, A. (ed.) (1987) *The Travel and Tourism Industry*, Oxford: Pergamon.

Hohl, A. and Tisdell, C.A. (1993) 'How useful are environmental safety standards in economics? The example of safe minimum standards for protection of species', *Biodiversity and Conservation*, 2, 2: 168–181.

Holloway, J.C. (1994) *The Business of Tourism*, 4th edn, London: Pitman.

Holloway, J.C. and Plant, R.V. (1988) *Marketing for Tourism*, London: Pitman.

Horwath Consulting (1994) *United Kingdom Hotel Industry 1994*, London: Horwath International.

Hotelling, H. (1949) 'The economics of public recreation', *The Prewitt Report*, Land and Recreation Planning Division, National Park Service, US Department of the Interior, Washington, DC.

Hough, D.E. and Kratz, C.G. (1983) 'Can "good" architecture meet the market test?', *Journal of Urban Economics*, 14: 40–54.

Hughes, H.L. (1981) 'A tourism tax: the cases for and against', *International Journal of Tourism Management*, 2, 3: 196–206.

Hunter, C. and Green, H. (1995) *Tourism and the Environment: A Sustainable Relationship?*, London: Routledge.

Hymer, S.H. (1976) *The International Operations of National Firms: A Study of Direct Investment*, Cambridge, Mass.: MIT Press.

Inskeep, E. (1991) *Tourism Planning: An Integrated and Sustainable Approach*, The Hague: Van Nostrand Reinhold.

Institute of Business Ethics (IBE) (1994) *Benefiting Business and the Environment*, London: IBE.

Instituto Español de Turismo (1980) 'La balanza de pagos turística de España en 1977', *Estudios Turísticos*, 65: 91–115.

—— (1983) 'Balanza de pagos turística de España: años 1979 y 1980', *Estudios Turísticos*, 77–78: 133–157.

International Chamber of Commerce (ICC) (1991) *ICC Business Charter for Sustainable Development: Principles of Environmental Management*, Paris: ICC.

International Hotels Environment Initiative (IHEI) (1993) *Environmental Management for Hotels: The Industry Guide to Best Practice*, Oxford: Butterworth Heinemann.

International Union for Conservation of Nature (IUCN) (1980) *World Conservation Strategy*, Gland, Switzerland: IUCN.

Iso-Ahola, S.E. (1982) 'Towards a social psychological theory of tourism motivation: a rejoinder', *Annals of Tourism Research*, 9, 2: 256–261.

Jafari, J. (1987) 'Tourism models: the sociocultural aspects', *Tourism Management*, 8, 2: 151–159.

Jappelli, T. and Pagano, M. (1988) 'Liquidity constrained households in an Italian cross-section', Centre for Economic Policy Research Discussion Paper no. 257.

—— (1989) 'Consumption and capital market imperfections: an international comparison', *American Economic Review*, 79: 1088–1105.

REFERENCES

Jefferson, A. (1990) 'Marketing in national tourist offices', in C.P. Cooper, (ed.) *Progress in Tourism, Recreation and Hospitality Management*, vol. 2, London: Belhaven.

Jefferson, A. and Lickorish, L. (1988) *Marketing Tourism: A Practical Guide*, Harlow: Longman.

Jeffrey, D. and Hubbard, N.J. (1988) 'Foreign tourism, the hotel industry and regional economic performance', *Regional Studies*, 22, 4: 319–329.

Jenner, P. and Smith, C. (1992) 'The tourism industry and the environment', Special Report 2453, London: Economist Intelligence Unit.

Johnson, P. and Ashworth, J. (1990) 'Modelling tourism demand: a summary review', *Leisure Studies*, 9, 2: 145–160.

Johnson, P. and Thomas, B. (1990) 'Measuring the local employment impact of a tourist attraction: an empirical study', *Regional Studies*, 24, 5: 395–403.

—— (1992a) *Choice and Demand in Tourism*, London: Mansell.

—— (1992b) *Perspectives on Tourism Policy*, London: Mansell.

Jong, H.W. de and Shepherd W.G. (eds) (1986) *Mainstreams in Industrial Organization*, Boston, Mass.: Kluwer.

Jundin, S. (1983) 'Barns uppfattning om konsumtion, sparande och arbete (Children's conceptions about consumption, saving and work)', Stockholm, The Stockholm School of Economics, EFI (doctoral dissertation, English summary).

Kahneman, D., Slovic, P. and Tversky, A. (1982) *Judgment under Uncertainty: Heuristics and Biases*, Cambridge: Cambridge University Press.

Kalecki, M. (1939) *Essays in the Theory of Economic Fluctuations*, London: Allen & Unwin.

Keane, M.J. (1997) 'Quality and pricing in tourism destinations', *Annals of Tourism Research*, 24, 1: 117–130.

Keogh, B. (1990) 'Public participation in community tourism planning', *Annals of Tourism Research*, 17, 3: 449–465.

Kent, P. (1990) 'People, places and priorities: opportunity sets and consumers' holiday choice', in G. Ashworth and B. Goodall (eds) *Marketing Tourism Places*, London: Routledge.

—— (1991) 'Understanding holiday choices', in M.T. Sinclair and M.J. Stabler (eds) *The Tourism Industry: An International Analysis*, Wallingford: CAB International.

Khan, H., Seng, C.F. and Cheong, W.K. (1990) 'Tourism multiplier effects in Singapore', *Annals of Tourism Research*, 17, 3: 408–418.

Kinnaird, V., Kothari, U. and Hall, D. (1994) 'Tourism: gender perspectives', in V. Kinnaird and D. Hall (eds) *Tourism: A Gender Analysis*, Chichester: John Wiley.

Kirchler, E. (1988) 'Household economic decision-making', in W.F. van Raaij, G.M. van Veldhoven and K-E. Wärneryd, *Handbook of Economic Psychology*, Dordrecht: Kluwer.

Kirker, C. (1994) 'Standardisation or specialistation: a happy medium?', Presentation to the Third Symposium on Tourism, Barcelona, 21 January.

Kirzner, I.M. (1973) *Competition and Entrepreneurship*, Chicago, Ill.: Chicago University Press.

Kliman, M.L. (1981) 'A quantitative analysis of Canadian overseas tourism', *Transportation Research*, 15A, 6: 487–497.

Kneese, A., Ayres, R. and d'Arge, R. (1970) *Economics and the Environment: A Materials Balance Approach*, Washington, DC: Resources for the Future.

Knudsen, O. and Parnes, A. (1975) *Trade Instability and Economic Development*, Lexington, Mass.: D.C. Heath.

Korca, P. (1991) 'Assessment of the environmental impacts of tourism', proceedings of an *International Symposium on the Architecture of Tourism in the Mediterranean*, Istanbul, Turkey: Yildiz University Press.

Kotler, P. (1991) *Marketing Management: Analysis, Planning, Implementation and Control*, London: Prentice Hall.

Kotler, P., Haider, D.H. and Rein, I. (1993) *Marketing Places*, New York: The Free Press.

Krippendorf, J. (1987) *The Holiday Makers*, London: Heinemann.

Krugman, P.R. (1980) 'Scale economies, product differentiation and the pattern of trade', *American Economic Review*, 70: 950–959.

—— (1989a) 'Industrial organization and international trade', in R. Schmalensee and R.D. Willig (eds) *Handbook of Industrial Organization*, Amsterdam: North Holland.

—— (1989b) 'New trade theory and the less developed countries', in G. Calvo and World Institute for Development Economics Research (eds) *Debt, Stabilization and Development*, Oxford: Basil Blackwell.

Lakatos, I. and Musgrave, A. (eds) (1970) *Criticism and the Growth of Knowledge*, Cambridge: Cambridge University Press.

Lancaster, K.J. (1966) 'A new approach to consumer theory', *Journal of Political Economy*, 84: 132–157.

Laws, E. (1991) *Tourism Marketing: Service and Quality Management Perspectives*, Cheltenham: Stanley Thornes.

Layard, R. (ed.) (1972) *Cost Benefit Analysis*, Harmondsworth: Penguin.

Lea, J. (1988) *Tourism and Development in the Third World*, London and New York: Routledge.

Lee, C-K., Var, T. and Blaine, T.W. (1996) 'Determinants of inbound tourist expenditures', *Annals of Tourism Research*, 23, 3: 527–542.

Lee, G. (1987) 'Tourism as a factor in development cooperation', *Tourism Management*, 8, 1: 2–19.

Lee, W. (1991) 'Prostitution and tourism in South-East Asia', in N. Redclift and M.T. Sinclair (eds) *Working Women: International Perspectives on Labour and Gender Ideology*, London and New York: Routledge.

Leiper, N. (1984) 'Tourism and leisure: the significance of tourism in the leisure spectrum', Proceedings of the twelfth New Zealand Geography Conference, Christchurch: New Zealand Geography Society.

Leontidou, L. (1994) 'Gender dimensions of tourism in Greece: employment, subcultures and restructuring', in V. Kinnaird and D. Hall (eds) *Tourism: A Gender Analysis*, Chichester: John Wiley.

Lerner, A.P. (1934) 'The concept of monopoly and the measurement of monopoly power', *Review of Economic Studies*, 1: 157–175.

Levine, M.E. (1987) 'Airline competition in deregulated markets: theory, firm strategy and public policy', *Yale Journal on Regulation*, 4: 393–494.

Lichfield, N. (1988) *Economics of Urban Conservation*, Cambridge: Cambridge University Press.

Liebenstein, H. (1950) 'Bandwagon, snob and Veblen Effects in the theory of consumers' demand', *Quarterly Journal of Economics*, 64, 2: 183–201.

Liebermann, M.B. and Montgomery, D.B. (1988) 'First-mover advantages', *Strategic Management Journal*, 9: 41–58.

Lin, T-b. and Sung, Y-W. (1983) 'Hong Kong', in E.A. Pye and T-b Lin (eds) *Tourism in Asia: The Economic Impact*, Singapore: Singapore University Press.

Lindberg, K. and Johnson, R.L. (1997) 'The economic values of tourism's social impacts', *Annals of Tourism Research*, 24, 1: 90–116.

Linder, S.B. (1961) *An Essay on Trade and Transformation*, London: John Wiley.

Liston, K. (1986) 'David and Goliath', *Courier*, November/December: 19–21.

Little, I.M.D. and Mirlees, J.A. (1974) *Project Appraisal and Planning for Developing Countries*, London: Heinemann.

Little, J.S. (1980) 'International travel in the UK balance of payments', *New England Economic Review*, May: 42–55.

Littlechild, S.C. (1986) *The Fallacy of the Mixed Economy*, 2nd edn, London: Institute of Economic Affairs.

Lockwood, M., Loomis, J. and DeLacy, T. (1993) 'A contingent valuation survey and benefit-cost analysis of forest preservation in East Gippsland, Australia', *Journal of Environmental Managment*, 38: 233–243.

Loeb, P.D. (1982) 'International travel to the United States: an econometric evaluation', *Annals of Tourism Research*, 9, 1: 7–20.

Loewenstein, G. (1987) 'Anticipation and the value of delayed consumption', *Economic Journal*, 97: 666–684.

Lombardi, P. and Sirchia, G. (1990) 'Il quarterre 16 IACF di Torino', in R. Roscelli (ed.) *Misurare Nell'Incertezza*, Turin: Celid.

Long, V.H. (1991) 'Government–industry–community interaction in tourism development in Mexico', in M.T. Sinclair and M.J. Stabler (eds) *The Tourism Industry: An International Analysis*, Wallingford: CAB International.

Long, V.H. and Kindon, S.L. (1997) 'Gender and tourism development in Balinese villages', in M.T. Sinclair (ed.) *Gender, Work and Tourism*, London and New York: Routledge.

Loomis, J.B., Creel, M. and Park, T. (1991) 'Comparing benefit estimates from travel cost and contingent valuation using confidence intervals from Hicksian welfare measures', *Applied Economics*, 23: 1725–1731.

Lozato, J.P. (1985) *Géographie du tourisme*, Paris: Masson.

Lucas Jr, R.E. (1972) 'Expectations and the neutrality of money', *Journal of Economic Theory*, 90: 103–124.

—— (1977) 'Understanding business cycles', in *Stabilization of the Domestic and International Economy*, Carnegie-Rochester Series on Public Policy, vol. 5: 7–30.

—— (1988) 'On the mechanics of economic growth', *Journal of Monetary Economics*, 22: 3–42.

Lundberg, D.E. (1989) *The Tourism Business*, New York: Van Nostrand Reinhold.

Lundberg, D.E., Krishnamoorthy, M. and Stavenga, M.H. (1995) *Tourism Economics*, New York: John Wiley.

Lundgren, J.O.J. (1982) 'The tourist frontier of Nouveau Quebec: functions and regional linkages', *Tourist Review*, 37, 2: 10–16.

Lyons, B. (1989) 'Barriers to entry', in S. Davies *et al.* (eds) *Economics of Industrial Organisation*, London and New York: Longman.

MaCurdy, T.E. (1982) 'The use of time-series processes to model the error structure of earnings in longitudinal data analysis', *Journal of Econometrics*, 18: 83–114.

Mak, J. and Nishimura, E. (1979) 'The economics of a hotel room tax', *Journal of Travel Research*, spring: 2–6.

Makridakis, S. (1986) 'The art and science of forecasting: an assessment and future directions', *International Journal of Forecasting*, 2, 1: 15–39.

March, J.G. (1962) 'The business firm as a political coalition', *Journal of Politics*, 24: 662–678.

March, J.G. and Simon, H.A. (1958) *Organizations*, New York: John Wiley.

Marglin, S.A. (1967) *Approaches to Dynamic Investment Planning*, Amsterdam: North Holland.

Martin, C.A. and Witt, S.F. (1987) 'Tourism demand forecasting models: choice of appropriate variable to represent tourists' cost of living', *Tourism Management*, 8, 3: 233–246.

—— (1988) 'Substitute prices in models of tourism demand', *Annals of Tourism Research*, 15, 2: 255–268.

—— (1989) 'Forecasting tourism demand: a comparison of the accuracy of several quantitative methods', *International Journal of Forecasting*, 5, 1: 1–13.

Martin, S. (1993) *Advanced Industrial Economics*, Cambridge, Mass.: Basil Blackwell.

Marx, K. (1967) *Capital* (centennial edition of *Das Kapital*, 1867), New York: International Publishers.

Maslow, A.H. (1954) *Motivation and Personality*, New York: Harper & Row.

—— (1968) *Towards a Psychology of Being*, 2nd edn, New York: Van Nostrand Reinhold.

Mason, E.S. (1957) *Economic Concentration and the Monopoly Problem*, Cambridge, Mass.: Harvard University Press.

Mathieson, A. and Wall, G. (1982) *Tourism: Economic, Physical and Social Impacts*, London: Longman.

McConnell, K.E. (1985) 'The economics of outdoor recreation', in A.V. Kneese and J.L. Sweeney (eds) *Handbook of Natural Resource and Energy Economics*, Amsterdam: North Holland, Elsevier Science.

McIntosh, R.W. and Goeldner, C.R. (1990) *Tourism Principles, Practices, Philosophy*, 6th edn, New York: John Wiley.

McVey, M. (1986) 'International hotel chains in Europe: survey of expansion plans as Europe is "rediscovered"', *Travel and Tourism Analyst*, September: 3–23.

Meadows, D.H., Meadows, D.L., Randers, J. and Behrens III, W.W. (1972) *The Limits of Growth: A Report for the Club of Rome's Project on the Predicament of Mankind*, London: Earth Island.

Means, G. and Avila, R. (1986) 'Econometric analysis and forecasts of US international travel: using the new TRAM model', *World Travel Overview, 1986/87*: 90–107.

—— (1987) 'An econometric analysis and forecast of US travel and the 1987 TRAM model update', *World Travel Overview, 1987/88*, 102–123.

Melville, J.A. (1995) 'Some empirical results for the airline and air transport markets of a small developing country', Ph.D. thesis, University of Kent at Canterbury.

Middleton, V.T.C. (1988) *Marketing in Travel and Tourism*, London: Heinemann.

Middleton, V.T.C. and Hawkins, R. (1993) 'Practical environmental policies in travel and tourism', *Travel and Tourism Analyst*, 6: 63–76, London: Economic Intelligence Unit.

Mill, R.C. and Morrison, A.M. (1985) *The Tourism System: An Introductory Text*, Englewood Cliffs, NJ: Prentice Hall.

Miossec, J.M. (1976) *Eléments pour une théorie de l'espace touristique*, Aix-en-Provence: Cahiers du Tourisme C–36, CHET.

Mishan, E.J. (1971) *Cost Benefit Analysis*, London: Allen & Unwin.

Mitchell, F. (1970) 'The value of tourism in East Africa', *East Africa Economic Review*, 2, 1: 1–21.

Mitchell, R.C. and Carson, R.T. (1989) *Using Surveys to Value Public Goods: The Contingent Valuation Method*, Washington, DC: Resources for the Future.

Mody, A. (1990) 'Institutions and dynamic comparative costs', *Cambridge Journal of Economics*, 14: 291–314.

Moeller, G.H. and Shafer, E.L. (1987) 'The Delphi technique: a tool for long-range tourism and travel planning', in J.R.B. Ritchie and C.R. Goeldner (eds) *Travel, Tourism and Hospitality Research*, New York: John Wiley.

Momsen, J. Henshall (1994) 'Tourism, gender and development in the Caribbean', in V. Kinnaird and D. Hall (eds) *Tourism: A Gender Analysis*, Chichester: John Wiley.

Monopolies and Mergers Commission (1989) *Thomson Travel Group and Horizon Travel Ltd.*, London: HMSO.

Moorhouse, J.C. and Smith, M.S. (1994) 'The market for residential architecture: 19th century row houses in Boston's South End', *Journal of Urban Economics*, 35: 267–277.

Mudambi, R. (1994) 'A Ricardian excursion to Bermuda: an estimation of mixed strategy equilibrium', *Applied Economics*, 26: 927–936.

Muroi, H. and Sasaki, N. (1997) 'Tourism and prostitution in Japan', in M.T. Sinclair (ed.) *Gender, Work and Tourism*, London and New York: Routledge.

Murphy, P.E. (1985a) *Tourism: A Community Approach*, New York: Methuen.

—— (1985b) 'Tourism and sustainable development', in W. Theobald (ed.) *Global Tourism: The Next Decade*, Oxford: Butterworth-Heinemann.

National Oceanic and Atmospheric Administration (NOAA) (1993) 'Report of the NOAA panel on contingent valuation', mimeo dated 12 January 1993, National Oceanic and Atmospheric Administration.

Nelson, C.R. and Plosser, C.I. (1982) 'Trends and random walks in macroeconomic time series', *Journal of Monetary Economics*, 10, 2: 139–162.

Nelson, R.R. and Winter, S.G. (1982) *An Evolutionary Theory of Economic Change*, Cambridge, Mass.: Harvard University Press.

Newbould, G. (1970) *Management and Merger Activity*, Liverpool: Guthstead.

Nijkamp, P. (1975) 'A multicriteria analysis for project evaluation: economic–ecological evaluation of a land reclamation project', *Papers of the Regional Science Association*, 35: 87–111.

—— (1988) 'Culture and region: a multidimensional evaluation of movements', *Environment and Planning B: Planning and Design*, 15: 5–14.

North, D.C. (1990) *Institutional Change and Economic Performance*, Cambridge: Cambridge University Press.

Norton, G.A. (1984) *Resource Economics*, London: Edward Arnold.

Obstfeld, M. (1990) 'Intertemporal dependence, impatience, and dynamics', *Journal of Monetary Economics*, 26: 45–75.

O'Hagan, J.W. and Harrison, M.J. (1984) 'Market shares of US tourist expenditure in Europe: an econometric analysis', *Applied Economics*, 16, 6: 919–931.

Oi, W.Y. and Hurter, A.P. (1965) *Economics of Private Truck Transportation*, Dubuque, Iowa: William C. Brown.

Oppermann, M. (1993) 'Tourism space in developing countries', *Annals of Tourism Research*, 20, 3: 535–556.

Opschoor, J.B. and Pearce, D.W. (eds) (1991) *Persistent Pollutants: Economics and Policy*, Dordrecht: Kluwer.

Opschoor, J.B. and Turner, R.K. (eds) (1993) *Environmental Economic and Policy Instruments: Principles and Practice*, Dordrecht: Kluwer.

Opschoor, J.B. and Vos, J. (1989) *Economic Instruments for Environmental Protection*, Paris: OECD.

Organization for Economic Co-operation and Development (OECD) (1981a) *The Impact of Tourism on the Environment*, Paris: OECD.

—— (1981b) *Case Studies of the Impact of Tourism on the Environment*, Paris: OECD.

O'Riordan, T. (1992) 'The precautionary principle in environmental management', GEC 92–103, CSERGE Working Paper, University of East Anglia and University College London.

Outbound Travel Industry Digest (1994) 4 February.

Pack, A. and Sinclair, M.T. (1995a) *Tourism, Conservation and Sustainable Development, Indonesia*, Report for the Overseas Development Administration, London.

—— (1995b) *Tourism, Conservation and Sustainable Development, India*, Report for the Overseas Development Administration, London.

Pack, A., Clewer, A. and Sinclair, M.T. (1995) 'Regional concentration and dispersal of tourism demand in the UK', *Regional Studies*, 29, 6: 570–576.

Paelinck, J.H.P. (1976) 'Qualitative multiple criteria analysis, environmental protection and multiregional development', *Papers of the Regional Science Association*, 36: 59–74.

240

REFERENCES

Page, S.J. (1993) 'Highlight on the Channel Tunnel', *Tourism Management*, 14, 6: 419–423.

—— (1994) *Transport for Tourism*, London and New York: Routledge.

Page, S.J. and Sinclair, M.T. (1989) 'Tourism and accommodation in London: alternative policies and the Docklands experience', *Built Environment*, 15, 2: 125–137.

—— (1992a) 'The Channel Tunnel and tourism markets in the 1990s', *Travel and Tourism Analyst*, February: 5–32.

—— (1992b) 'The Channel Tunnel: an opportunity for London's tourism industry?', *Tourism Recreation Research*, 17, 2: 57–70.

Pattison, T. (1992) *The Future for the Coach Industry*, Insights, no. 5 London: English Tourist Board.

Pawson, I.G., Stanford, D.D., Adams, V.A. and Nurbu, M. (1984) 'Growth of tourism in Nepal's Everest region: impact on the physical environment and structure of human settlements', *Mountain Research and Development*, 4, 3: 237–246.

Pearce, D.G. (1987) *Tourism Today: A Geographical Analysis*, Harlow: Longman.

—— (1989) *Tourist Development*, 2nd edn, Harlow: Longman.

Pearce, D.G. and Butler, R.W. (eds) (1993) *Tourism Research: Critiques and Challenges*, London: Routledge.

Pearce, D.W. (1976) *Environmental Economics*, London: Longman.

—— (1993) 'Sustainable development', in D.W. Pearce (ed.) *Ecological Economics: Essays in the Theory and Practice of Environmental Economics*, London: Edward Elgar.

Pearce, D.W. and Nash, C.A. (1981) *The Social Appraisal of Projects*, London: Macmillan.

Pearce, D.W. and Turner, R.K. (1990) *Economics of Natural Resources and the Environment*, London and New York: Harvester Wheatsheaf.

Pearce, D.W., Markandya, A. and Barbier, E.B. (1989) *Blueprint for a Green Economy*, London: Earthscan.

Pearce, P.L. (1982) *The Social Psychology of Tourist Behaviour*, Oxford: Pergamon.

Peltzman, S. (1977) 'The gains and losses from industrial concentration', *Journal of Law and Economics*, 20: 229–263.

Pischke, J-S. (1991) 'Individual income, incomplete information and aggregate consumption', Industrial Relations Section working paper no. 289, Princeton University, NJ.

Pizam, A. and Calantone, R. (1987) 'Beyond psychographics: values as determinants of tourist behaviour', *International Journal of Hospitality Management*, 6, 3: 177–181.

Plog, S.C. (1973) 'Why destination areas rise and fall in popularity', *Cornell HRA Quarterly*, November: 13–16.

—— (1987) 'Understanding psychographics in tourism research', in J.R. Brent Ritchie and C.R. Goeldner (eds) *Travel, Tourism and Hospitality Research: A Handbook for Managers and Researchers*, New York: John Wiley.

Pollard, H.J. (1976) 'Antigua, West Indies: an example of the operation of the multiplier process arising from tourism', *Revue de Tourisme*, 3: 30–34.

Poon, A. (1988) 'Innovation and the future of Caribbean tourism', *Tourism Management*, 9, 3: 213–220.

Posner, M.W. (1961) 'International trade and technical change', *Oxford Economic Papers*, 13: 323–341.

Prais, S.J. (1976) *The Evolution of Giant Firms in Britain*, National Institute of Economic and Social Research, Economic and Social Studies, 30, Cambridge: Cambridge University Press.

Puhipan (1994) 'Boycott paradise', *In Focus*, 12, summer: 10–11.

Purcell, K. (1997) 'Women's employment in UK tourism: gender roles and labour markets', in M.T. Sinclair (ed.) *Gender, Work and Tourism*, London and New York: Routledge.

Pyo, S.S., Uysal, M. and McLellan, R.W. (1991) 'A linear expenditure model for tourism demand', *Annals of Tourism Research*, 18: 443–454.

Qualls, W.J. (1982) 'Changing sex roles: its impact upon family decision making', *Advances in Consumer Research*, 9: 151–162.

Quayson, J. and Var, J. (1982) 'A tourism demand function for the Okanagan, BC', *Tourism Management*, 3, 2: 108–115.

Randall, A. (1993) 'The problem of market failure', in R. Dorfman and N.S. Dorfman (eds) *Economics of the Environment*, 3rd edn, New York: Norton.

Randall, J. (1986) 'European airlines move into hotels. survey of the leading carriers' expanding hotel interests', *Travel and Tourism Analyst*, July: 45–54.

Rao, A. (1986) *Tourism and Export Instability in Fiji*, Occasional Papers in Economic Development no. 2, Faculty of Economic Studies, University of New England, Australia.

Redefining Progress (1995) *The Genuine Progress Indicator: Summary of Data and Methodology*, San Francisco: Redefining Progress.

Reekie, W.D. (1984) *Markets, Entrepreneurs and Liberty: An Austrian View of Capitalism*, Brighton: Wheatsheaf.

Reinganum, J. (1989) 'The timing of innovation: research development and diffusion', in R. Schmalensee and R.D. Willig (eds) *Handbook of Industrial Organization*, vol. 1, Amsterdam: North Holland.

Richardson, H.W. (1972) *Input–Output and Regional Economics*, London: Weidenfeld & Nicolson.

Ritchie, J.R.B. and Goeldner, C.R. (eds) (1987) *Travel, Tourism and Hospitality Research: A Handbook for Managers and Researchers*, New York: John Wiley.

Rivera-Batiz, L.A. and Romer, P.M. (1991) 'Economic integration and endogenous growth', *Quarterly Journal of Economics*, 106: 531–555.

Robinson, H. (1976) *A Geography of Tourism*, London: MacDonald & Evans.

Rodrik, D. (1995) 'Getting intervention right: how South Korea and Taiwan grew rich', *Economic Policy*, 20: 53–107.

Romer, P.M. (1986) 'Increasing returns and long-run growth', *Journal of Political Economy*, 94: 1002–1037.

—— (1994) 'The origins of endogenous growth', *Journal of Economic Perspectives*, 8: 3–22.

Romeril, M. (1989) 'Tourism and the environment: accord or discord', *Tourism Management*, 10, 3: 204–208.

Roscelli, R. and Zorzi, F. (1990) 'Valutazione di progetti di riqualificazione urbana', in R. Roscelli (ed.) *Misurare Nell'Incertezza*, Turin: Celid.

Rosen, S. (1974) 'Hedonic prices and implicit markets: production differentiation in pure competition', *Journal of Political Economy*, 82, 1: 34–55.

Rosenberg, M. (1956) 'Cognitive structure and attitudinal effect', *Journal of Abnormal and Social Psychology*, 53: 367–372.

Ryan, C. (1991) 'UK package holiday industry', *Tourism Management*, 12, 1: 76–77.

Saaty, R.W. (1987) 'The analytic hierarchy process: what it is and how it is used', *Mathematical Modelling*, 9: 161–176.

Sadler, P., Archer, B.H. and Owen, C. (1973) *Regional Income Multipliers*, Occasional Papers in Economics, no. 1, Bangor: University of Wales Press.

Sakai, M.Y. (1988) 'A micro-analysis of business travel demand', *Applied Economics*, 20: 1481–1496.

Salop, S.C. (1979a) 'Monopolistic competition with outside goods', *Bell Journal of Economics*, 10: 141–156.

—— (1979b) 'Strategic entry deterrence', *American Economic Review Papers and Proceedings*, 69: 335–338.

Samuelson, P. (1948) 'International trade and the equalization of factor prices', *Economic Journal*, 58: 163–184.

—— (1949) 'International factor price equalization once again', *Economic Journal*, 59: 181–197.

Sargent, T. and Wallace, N. (1976) 'Rational expectations and the theory of economic policy', *Journal of Monetary Economics*, 2: 169–183.

Schaeffer, P.V. and Millerick, C.A. (1991) 'The impact of historic district designation on property values: an empirical study', *Economic Development Quarterly*, 5, 4: 301–312.

Scherer, F.M. (1967) 'Research and development resource allocation under rivalry', *Quarterly Journal of Economics*, 81: 359–394.

—— (1970) *Industrial Pricing: Theory and Evidence*, Chicago, Ill.: Rand McNally.

Schmalensee, R. (1972) *The Economics of Advertising*, Amsterdam: North Holland.

Schmalensee, R. and Willig, R.D. (eds) (1989) *Handbook of Industrial Organization*, vols 1 and 2, Amsterdam: North Holland.

Schmoll, G.A. (1977) *Tourism Promotion*, London: Tourism International Press.

Schumacher, E.F. (1973) *Small is Beautiful: A Study of Economics as if People Mattered*, London: Blond & Briggs.

Schwaninger, M. (1989) 'Trends in leisure and tourism for 2000 to 2010: scenario with consequences for planners', in S.F. Witt and L. Moutinho (eds) *Tourism Marketing and Management Handbook*, Hemel Hempstead: Prentice Hall.

Scott, A.D. (1955) 'The fishery: the objective of sole ownership', *Journal of Political Economy*, 63: 124–142.

Scott, J. (1997) 'Chances and choices: women and tourism in Northern Cyprus', in M.T. Sinclair (ed.) *Gender, Work and Tourism*, London and New York: Routledge.

Scottish Tourist Board (1993) *Going Green: Guideline for the Scottish Tourism Industry*, Edinburgh: Scottish Tourist Board.

Seaton, A.V. (ed.) (1994) *Tourism: The State of the Art*, Chichester: John Wiley.

Seely, R.L., Iglarsh, H.J. and Edgell, D.L. (1980) 'Utilizing the Delphi technique at international conferences: a method for forecasting international tourism conditions', *Travel Research Journal*, 1: 30–35.

Sen, A. (1979) 'Rational fools', in F. Hahn and M. Hollis (eds) *Philosophy and Economic Theory*, Oxford: Oxford University Press.

Sessa, A. (1983) *Elements of Tourism Economics*, Rome: Catal.

—— (1984) 'Comments on Peter Gray's "The contribution of economics to tourism"', *Annals of Tourism Research*, 11: 283–302.

Shafer, E.L., Moeller, G.H. and Getty, R.E. (1974) *Future Leisure Environment*, Forest Research Paper NE-301, USDA Forest Experiment Station, Pennsylvania.

Shaw, G. and Williams, A.M. (1994) *Critical Issues in Tourism*, Oxford: Basil Blackwell.

Shaw, S. (1987) *Airline Marketing and Management*, London: Pitman.

Shelby, B. and Heberlein, T.A. (1984) 'A conceptual framework for carrying capacity determination', *Leisure Sciences*, 6, 4: 433–451.

Sheldon, P.J. (1986) 'The tour operator industry: an analysis', *Annals of Tourism Research*, 13, 3: 349–365.

—— (1990) 'A review of tourism expenditure research', in C.P. Cooper (ed.) *Progress in Tourism, Recreation and Hospitality Management*, vol. 2, London: Belhaven.

—— (1994) 'Tour operators', in S.F. Witt and L. Moutinho (eds) *Tourism Management and Marketing Handbook*, 2nd edn, Hemel Hempstead: Prentice Hall.

Simon, H.A. (1955) 'A behavioural model of rational choice', *Quarterly Journal of Economics*, 69: 99–118.

—— (1957) *Models of Man*, New York: Wiley.

—— (1979) 'Rational decision making in business organizations', *American Economic Review*, 69: 493–514.

Sinclair, M.T. (1990) *Tourism Development in Kenya*, Washington, DC: World Bank.

—— (1991a) 'The economics of tourism', in C.P. Cooper (ed.) *Progress in Tourism, Recreation and Hospitality Management*, vol. 3, London: Belhaven.

—— (1991b) 'The tourism industry and foreign exchange leakages in a developing country', in M.T. Sinclair and M.J. Stabler (eds) *The Tourism Industry: An International Analysis*, Wallingford: CAB International.

—— (1991c) 'Women, work and skill: economic theories and feminist perspectives, in N. Redclift and M.T. Sinclair (eds) *Working Women: International Perspectives on Labour and Gender Ideology*, London and New York: Routledge.

—— (1992a) 'Tourism, economic development and the environment: problems and policies', in C.P. Cooper and A. Lockwood (eds) *Progress in Tourism, Recreation and Hospitality Management*, vol. 4, London: Belhaven.

—— (1992b) 'Tour operators and tourism development policies in Kenya', *Annals of Tourism Research*, 19, 3: 555–558.

—— (1997a) *Tourism and Economic Development: A Survey*, Studies in Economics, 97/3, University of Kent at Canterbury.

—— (ed.) (1997b) *Gender, Work and Tourism*, London and New York: Routledge.

—— (1997c) 'Issues and theories of gender and work in tourism', in M.T. Sinclair (ed.) *Gender, Work and Tourism*, London and New York: Routledge.

Sinclair, M.T. and Bote Gómez, V. (1996) 'Tourism, the Spanish economy and the balance of payments', in M. Barke, M. Newton and J. Towner (eds) *Tourism in Spain: Critical Perspectives*, Wallingford: CAB International

Sinclair, M.T. and Page, S.J. (1993) 'The Euroregion: a new framework for tourism and regional development', *Regional Studies*, 27, 5: 475–483.

Sinclair, M.T. and Stabler, M.J. (eds) (1991) *The Tourism Industry: An International Analysis*, Wallingford: CAB International.

Sinclair, M.T. and Sutcliffe, C.M.S. (1978) 'The first round of the Keynesian income multiplier', *Scottish Journal of Political Economy*, 25, 2: 177–186.

—— (1988a) 'The estimation of Keynesian income multipliers at the sub-national level', *Applied Economics*, 20, 11: 1435–1444.

—— (1988b) 'Negative multipliers: a case for disaggregated estimation, *Tijdschrift Voor Economische en Sociale Geografie*, 79, 2: 104–107.

—— (1989a) 'Truncated income multipliers and local income generation over time', *Applied Economics*, 21, 12: 1621–1630.

—— (1989b) 'The economic effects on destination areas of foreign involvement in the tourism industry: a Spanish application', in B. Goodall and G. Ashworth (eds) *Marketing in the Tourism Industry: The Promotion of Destination Regions*, Beckenham: Croom Helm.

Sinclair, M.T. and Tsegaye, A. (1990) 'International tourism and export instability', *Journal of Development Studies*, 26, 3: 487–504.

Sinclair, M.T. and Vokes, R. (1992) 'The economics of tourism in Asia and the Pacific', in M. Hitchcock, V.T. King and M. Parnwell (eds) *Tourism in South-East Asia: Theory and Practice*, London and New York: Routledge.

Sinclair, M.T., Alizadeh, P. and Atieno Adero Onunga, E. (1992) 'The structure of international tourism and tourism development in Kenya', in D. Harrison (ed.) *Tourism and the Less Developed Countries*, London: Belhaven.

Sinclair, M.T., Clewer, A. and Pack, A. (1990) 'Hedonic prices and the marketing of package holidays', in G. Ashworth and B. Goodall (eds) *Marketing Tourism Places*, London and New York: Routledge.

—— (1994) 'Estrategias del turismo metropolitano: el caso de Londres', *Estudios Turísticos*, 124: 15–30.

Sindiyo, D.M. and Pertet, F.N. (1984) 'Tourism and its impact on wildlife in Kenya', *Industry and Environment*, 7, 1: 14–19.

Slovic, P, Fischoff, B. and Lichtenstein, S. (1977) 'Behavioural decision theory', *Annual Review of Psychology*, 28: 1–39.

Smeral, E. (1988) 'Tourism demand, economic theory and econometrics: an integrated approach', *Journal of Travel Research*, 26, 4: 38–43.

Smeral, E. and Witt, S.F. (1996) 'Econometric forecasts of tourism demand to 2005', *Annals of Tourism Research*, 23, 4: 891–907.

Smith, A. (1994) 'Imperfect competition and international trade', in D. Greenaway and L.A. Winters (eds) *Surveys in International Trade*, Oxford: Basil Blackwell.

Smith, C. and Jenner, P. (1984) 'Tourism and the environment', *Travel and Tourism Analyst*, 5: 68–86.

Smith, S.L.J. (1983) *Recreation Geography*, London: Longman.

—— (1989) *Tourism Analysis: A Handbook*, Harlow: Longman.

Smith, V.K. and Desvouges, W. (1986) *Measuring Water Quality Benefits*, Boston, Mass.: Kluwer.

Smith, V.K., Palmquist, R.B. and Jakus, P. (1991) 'Combining Farrel frontier and hedonic travel cost models for valuing estuarine quality', *Review of Economics and Statistics*, 63, 4: 694–699.

Smith, V.L. (1968) 'Economics of production from natural resources', *American Economic Review*, 58, 3: 409–431.

—— (ed.) (1989) *Hosts and Guests: The Anthropology of Tourism*, Philadelphia: University of Pennsylvania Press.

Socher, K. (1986) 'Tourism in the theory of international trade and payments', *The Tourist Review*, 3: 24–26.

Solow, R.M. (1956) 'A contribution to the theory of economic growth', *Quarterly Journal of Economics*, 70: 65–94.

Song, B-N and Ahn, C-Y. (1983) 'Korea', in E.A. Pye and T-b. Lin (eds) *Tourism in Asia: The Economic Impact*, Singapore: Singapore University Press.

Spence, A.M. (1977) 'Entry, capacity, investment and oligopolistic pricing', *Bell Journal of Economics*, 8: 534–544.

Stabler, M.J. (1995a) 'Research in progress on the economic and social value of conservation' in P. Burman, R. Pickard and S. Taylor (eds) *The Economics of Architectural Conservation*, York: Institute of Advanced Architectural Studies, University of York.

—— (1995b) 'Sustainability? What is in it for us? An economic view of how to encourage heritage and tourism enterprises to change their management strategies and methods to attain environmental goals', Paper presented at the Heritage Interpretation Fourth International Global Conference, Barcelona, March.

—— (1996a) 'The emerging new world of leisure quality: does it matter and can it be measured?', in M. Collins (ed.) *Leisure in Different Words*, vol. 2, *Leisure in Industrial and Post-Industrial Societies*, Eastbourne: Leisure Studies Association.

—— (1996b) 'Managing the leisure natural resource base: utter confusion or evolving consensus?' Paper presented at the World Leisure and Recreation Association Fourth World Congress, Free Time and the Quality of Life for the 21st Century, Cardiff, July.

—— (1996c) 'The role of land-use planning in sustaining tourism natural resources: an economic perspective', Paper presented at the Sixth International Symposium on Society and Resource Management: Social Behaviour, Natural Resources and the Environment, Pennsylvania State University, May.

Stabler, M.J. and Goodall, B. (1992) 'Environmental auditing in the quest for sustainable tourism: the destination perspective', Papers and proceedings of conference, Tourism in Europe, University of Durham, July.

—— (1996) 'Environmental auditing in planning for sustainable island tourism', in L. Briguglio, B. Archer, J. Jafari and G. Wall *Sustainable Tourism in Islands and Small States: Issues and Policies*, London: Pinter (Cassell).

—— (1997) 'Environmental awareness, action and performance in the tourism industry: a case study of the hospitality sector in Guernsey', *Tourism Managment*, 18, 1: 19–33.

Stacey, B.G. (1982) 'Economic socialization in the pre-adult years', *British Journal of Social Psychology*, 21: 159–173.

Stavins, R. (ed.) (1988) *Project 88: Harnessing Market Forces to Protect Our Environment*, Public Policy Study sponsored by Senators Wirth and Heinz, Washington, DC.

Stiglitz, J.E. (1989) 'Imperfect information in the product market', in R. Schmalensee and R.D. Willig (eds) *Handbook of Industrial Organization*, vol. 1, Amsterdam: North Holland.

Stoneman, P. (1983) *The Economic Analysis of Technological Change*, Oxford: Oxford University Press.

Stronge, W.B. and Redman, M. (1982) 'US tourism in Mexico: an empirical analysis', *Annals of Tourism Research*, 9, 1: 21–35.

Sutcliffe, C.M.S. and Sinclair, M.T. (1980) 'The measurement of seasonality within the tourist industry: an application of tourist arrivals in Spain', *Applied Economics*, 12, 4: 429–441.

Swain, M. Byrne (ed.) (1995) 'Gender in Tourism', special issue, *Annals of Tourism Research*, 22, 2.

Syriopoulos, T. (1995) 'A dynamic model of demand for Mediterranean tourism', *International Review of Applied Economics*, 9, 3: 318–336.

Syriopoulos, T. and Sinclair, M.T. (1993) 'An econometric study of tourism demand: the AIDS model of US and European tourism in Mediterranean countries', *Applied Economics*, 25, 12: 1541–1552.

Tan, L. (1992) 'A Heckscher-Ohlin approach to changing comparative advantage in Singapore's manufacturing sector', *Weltwirtschaftliches Archiv*, 128: 288–309.

Taylor, P. (1997) 'Mixed strategy pricing behaviour in the UK package tour industry', *International Journal of the Economics of Business*, 4: 3.

Telfer, D.J. and Wall, G. (1996) 'Linkages between tourism and food production', *Annals of Tourism Research*, 23, 3: 635–653.

Teye, V.B. (1988) 'Prospects for regional tourism cooperation in Africa', *Tourism Management*, 9, 3: 221–234.

Theuns, H.L. (1991) *Third World Tourism Research 1950–1984: A Guide to the Literature*, Frankfurt: Peter Lang.

Thurot, J.M. (1980) *Capacité de chargé et production touristique*, Etudes et Memoires, 43, Centre de hautes etudes touristiques, Aix-en-Provence.

Tietenberg, T. (1988) *Environmental and Natural Resource Economics*, 2nd edition, Glenview, Ill.: Scott, Foresman.

Tirole, J. (1988) *The Theory of Industrial Organization*, Cambridge, Mass.: MIT Press.

Tisdell, C.A., Aislabie, C.J. and Stanton, P.J. (eds) (1988) *Economics of Tourism: Case Study and Analysis*, University of Newcastle, New South Wales: Institute of Industrial Economics.

Tooman, L.A. (1997) 'Applications of the life-cycle model in tourism', *Annals of Tourism Research*, 24, 1: 214–234.

Tourism Concern (1995) 'Our holidays, their homes', special issue on people displaced by tourism, *In Focus*, 15, spring: 3–13.

Travel Trade Gazette (1994) 2 March.

Tremblay, P. (1989) 'Pooling international tourism in Western Europe', *Annals of Tourism Research*, 16, 4: 477–491.

Tribe, J. (1995) *The Economics of Leisure and Tourism*, Oxford: Butterworth-Heinemann.

Troyer, W. (1992) *The Green Partnership Guide*, Toronto: Canadian Pacific Hotels and Resorts.

Turner, R.K. (ed.) (1988) *Sustainable Environmental Management: Principles and Practice*, London: Belhaven.

Turner, R.K., Pearce, D.W. and Bateman, I. (1994) *Environmental Economics: An Elementary Introduction*, London: Harvester Wheatsheaf.

Tyler, C. (1989) 'A phenomenal explosion', *Geographical Magazine*, 61, 8: 18–21.

Ungson, G.R., Braunstein, D.N. and Hall, P.D. (1981) 'Managerial information processing: a research review', *Administrative Science Quarterly*, 26: 116–134.

United Nations Conference on Trade and Development (UNCTAD) (1973) *Elements of Tourism Policy in Developing Countries*, Report by the Secretariat of UNCTAD, TD/B/C.3/89, Add.3, Geneva: UNCTAD.

—— (UNCTAD) (1988) *Trade and Development Report*, Geneva: UNCTAD.

Uysal, M. and Crompton, J.L. (1984) 'Determinants of demand for international tourist flows in Turkey', *Tourism Management*, 5, 4: 288–297.

Van der Ploeg, F. and Tang, P. (1994) 'Growth, deficits and research and development in the global economy', in F. van der Ploeg (ed.) *The Handbook of International Macroeconomics*, Oxford: Basil Blackwell.

Van Doorn, J.W.M. (1984) 'Tourism forecasting and the policymaker: criteria of usefulness', *Tourism Management*, 5, 1: 24–39.

Van Doren, C.S., Koh, Y.K. and McCahill, A. (1994) 'Tourism research: a state of the art citation analysis', in A.V. Seaton (ed.) *Tourism: The State of the Art*, Chichester: John Wiley.

Varley, R.C.G. (1978), *Tourism in Fiji: Some Economic and Social Problems*, Occasional Papers in Economics, no. 12, Bangor: University of Wales Press.

Vaughan, D.R. and Long, J. (1982) 'Tourism as a generator of employment: a preliminary appraisal of the position in Great Britain', *Journal of Travel Research*, 21, 2: 27–31.

Veblen, T. (1899) *The Theory of the Leisure Class*, New York: Mentor.

Vellas, F. (1989) 'Tourisme et economie internationale', *Teoros*, 7, 5: 36–39.

Vernon, R. (1966) 'International investment and international trade in the product cycle', *Quarterly Journal of Economics*, 80: 190–207.

Vickerman, R.W. (1993) 'Tourist implications of new transport opportunities: the Channel Tunnel', in S. Glyptis (ed.) *Leisure and the Environment*, London: Belhaven.

Voogd, H. (1988) 'Multicriteria evaluation: measures, manipulation and meaning: a reply', *Environment and Planning B: Planning and Design*, 15, 1: 65–72.

Wahab, S.E.A. (1975) *Tourism Management*, London: Tourism International Press.

Walsh, R.G. (1986) 'Recreation economic decisions: Comparing benefits and costs', State College, Pa.: Venture.

Wandner, S.A. and Van Erden, J.D. (1980) 'Estimating the demand for international tourism using time series analysis', in D.E. Hawkins, E.L. Shafer and J.M. Rovelstad (eds) *Tourism Planning and Development Issues*, Washington, DC: George Washington University.

Wanhill, S.R.C. (1980) 'Charging for congestion at tourist attractions', *International Journal of Tourism Management*, 1, 3: 168–174.

Wanhill, S.R.C. (1982) 'Evaluating the resource costs of tourism', *Tourism Management*, 3, 4: 208–211.

—— (1986) 'Which investment incentives for tourism?', *Tourism Management*, 7, 1: 2–7.

—— (1988) 'Tourism multipliers under capacity constraints', *Service Industries Journal*, 8: 136–142.

Welford, R. and Gouldson, A. (1993) *Environmental Management and Business Strategy*, London: Pitman.

Weston, R. (1983) 'The ubiquity of room taxes', *Tourism Management*, 4, 3: 194–198.

Wheeller, B. (1994) 'Ecotourism, sustainable tourism and the environment – a symbiotic or shambolic relationship', in A.V. Seaton (ed.) *Tourism: The State of the Art*, Chichester: John Wiley.

247

White, K.J. (1982) 'The demand for international travel: a system-wide analysis for US travel to Western Europe', Discussion Paper no. 82–28, University of British Columbia, Canada.

White, K.J. and Walker, M.B. (1982) 'Trouble in the travel account', *Annals of Tourism Research*, 9, 1: 37–56.

Wight, P. (1993) 'Ecotourism: ethics or eco-sell?', *Journal of Travel Research*, 31: 3–9.

—— (1994) 'The greening of the hospitality industry: economic and environmental good sense', in A.V. Seaton (ed.) *Tourism: The State of the Art*, Chichester: John Wiley.

Williams, A.M. and Shaw, G. (1988) 'Tourism: candy floss industry or job generator?', *Town Planning Review*, 59: 81–104.

Williamson, O.D. (1989) 'Transaction cost economics', in R. Schmalensee and R.D. Willig (eds) *Handbook of Industrial Organization*, vol. 1, Amsterdam: North Holland.

Williamson, O.E. (1985) *The Economic Institutions of Capitalism*, New York: Free Press.

—— (1986) *Economic Organization: Firms, Markets and Policy Control*, Brighton: Wheatsheaf.

Willis, K.G. (1989) 'Option value and non-user benefits of wildlife conservation', *Journal of Rural Studies*, 5, 3: 245–256.

Willis, K.G. and Benson, J.F. (1988) 'A comparison of user benefits and costs of nature conservation at three nature reserves', *Regional Studies*, 22, 5: 417–428.

Willis, K.G. and Garrod, G. (1991a) 'An individual travel cost method of evaluating forest recreation', *Journal of Agricultural Economics*, 42: 33–42.

—— (1991b) 'Valuing open access recreation on inland waterways: on-site recreational surveys and selection effects', *Regional Studies*, 25, 6: 511–524.

—— (1991c) 'Landscape values: a contingent valuation approach and case study of the Yorkshire Dales National Park', *Countryside Change Unit Working Paper 21*, Newcastle-upon-Tyne: University of Newcastle.

—— (1993a) 'The value of waterside properties: estimating the impact of waterways and canals on property values through hedonic price models and contingent valuation methods', *Countryside Change Unit Working Paper 44*, Newcastle-upon-Tyne: University of Newcastle.

—— (1993b) 'Valuing wildlife: the benefits of wildlife trusts', *Countryside Change Unit Working Paper 46*, Newcastle upon Tyne: University of Newcastle.

Willis, K.G., Garrod, G. and Dobbs, I.M. (1990) 'The value of canals as a public good: the case of the Montgomery and Lancaster Canals', *Countryside Change Unit Working Paper 5*, Newcastle-upon-Tyne: University of Newcastle.

Willis, K.G., Garrod, G., Saunders, C. and Whitby, M. (1993) 'Assessing methodologies to value the benefits of environmentally sensitive areas', *Countryside Change Unit Working Paper 39*, Newcastle upon Tyne: University of Newcastle.

Witt, C.A., Witt, S.F. and Wilson, N. (1994) 'Forecasting international tourist flows', *Annals of Tourism Research*, 21, 3: 612–628.

Witt, S.F. (1980) 'An econometric comparison of UK and German foreign holiday behaviour', *Managerial and Decision Economics*, 1, 3: 123–131.

Witt, S.F. and Martin, C.A. (1987) 'Econometric models for forecasting international tourism demand', *Journal of Travel Research*, 25, 3: 23–30.

—— (1989) 'Demand forecasting in tourism and recreation', in C.P. Cooper (ed.) *Progress in Tourism, Recreation and Hospitality Management*, vol. 1, London: Belhaven.

Witt, S.F. and Moutinho, L. (1994) *Tourism Management and Marketing Handbook*, 2nd edn, Hemel Hempstead: Prentice Hall.

World Commission on Environment and Development (WCED – the Brundtland Commission) (1987) *Our Common Future*, Oxford: Oxford University Press.

World Tourism Organization (WTO) (1992) *Tourism Trends to the Year 2000 and Beyond*, Madrid: World Tourism Organization.

REFERENCES

World Travel and Tourism Council (WTTC) (1994) *Green Globe: An Invitation to Join*, London: World Travel and Tourism Council.

Yokeno, N. (1968) 'La localisation de l'industrie touristique: application de l'analyse de Thunen-Weber', *Les Cahiers du Tourisme*, vol. C-9, Aix-en-Provence.

Young, A. (1991) 'Learning by doing and the dynamic effects of international trade', *Quarterly Journal of Economics*, 106: 369–405.

Zahedi, F. (1986) 'The analytic hierarchy process: a survey of the method and its application', *Interfaces*, 16: 96–108.

Zeldes, S.P. (1989) 'Consumption and liquidity constraints: an empirical investigation', *Journal of Political Economy*, 97: 305–46.

Zhou, D., Yanagida, J.F., Chakrovorty, U. and Leung, P. (1997) 'Estimating economic impacts from tourism', *Annals of Tourism Research*, 24, 1: 76–89.

NAME INDEX

SUBJECT INDEX